THE

"Four Cosmic Pillars"

In Colour

ISBN-13: 978- 1537680149

ISBN-10: 10 1537680145

WRITTEN BY PEET SCHUTTE

© KOSMOLOGIESE EN ASTRONOMIESE TEGNIKA

SINGULARITY

as APPLYING IN PHYSICS

I am P.S.J Schutte, nicknamed Peet. Being a white South African my mother tongue is Afrikaans and my second language is English. Because of a shortage of funding my books were never edited in any way. Where you do find language issues please rather think of this mistake in terms of my poverty than think of my stupidity. I do find much pride in my status as being Afrikaner and would like to have my names used by pronouncing it in the manner Afrikaans dictates...therefore I would sincerely appreciate the courtesy when readers will take note that my name and last name are pronounced in Afrikaans, which is originally from Dutch and must be pronounced that way. Peet one would pronounce "here" which is the closest English to the pronouncing of the "ee". The "Sch" in Schutte is pronounced exactly as school is where both actually are pronounced Skutte or "skool".

By pronouncing my name in Afrikaans you do me the utmost courtesy any one can. Being an Afrikaner is what I am most proud of. I submit article to well known physics magazines but my articles are rejected on the most unappeasable grounds and for the most outrageously ridiculous reasons the Newtonians can think of. I explain how gravity forms but I am rejected because they are of the opinion that my work does not meet. One such an article I may use because I said I was going to use the material as an open letter I gladly show.

Email as follows: orders@sirnewtonsfruad.com

WHOM IT MAY CONCERN,

I do find much pride in my status as being Afrikaner and would like to have my names used by pronouncing it in the manner Afrikaans dictates...therefore I would sincerely appreciate the courtesy when readers will take note that my name and last name are pronounced in Afrikaans, which is originally from Dutch and must be pronounced that way. Peet one would pronounce "here" which is the closest English to the pronouncing of the "ee". The "Sch" in Schutte is pronounced exactly as school is where both actually are pronounced Skutte or "skool". By pronouncing my name in Afrikaans you do me the utmost courtesy any one can. Being an Afrikaner is what I am most proud of. Another point I wish to highlight is that I feel compiled to produce this work in a comic-like format. I have found that the more intellectual and the more educated Academics are, the less they understand the most primitive or classical mistakes in science as well as physics.

As I said my mother tongue is Afrikaans and my second language is English. I have per suiting this theory that I partly present in this book, of which the investigating research was done the past thirty years. Then I compiled my presentation thereof for the past nine years on full time basis whereby I was tying to introduce my findings to many academics without much joy. This past nine years saw me go without any income as I tried to get my theorem recognised. Going without a steady income left me almost destitute and in order to find a manner to get my theory across to the attention of influential readers, I decided to publish these books electronically as to try and get around the stranglehold of Newtonian bias controlling science at present worldwide. I decided to publish these articles through LULU.com which I saw as way the only manner whereby I could generate funding by which I would be able to have the twenty seven books I already wrote linguistically edited and then to have the books published on a Print-On-Demand basis. With my first language not being English and the books not linguistically checked by an expert there are bound to be language errors that readers will notice. In the past I tried to check my work myself but after checking say one hundred and fifty pages for language corrections, instead of having corrected work I ended instead having four hundred pages of new written information which is still not language corrected but holds a lot more information. This is because my priorities lie elsewhere. I aim to spend money on correcting the work as far as language goes, as I receive money and in the hope that I will receive money. I will have all my work including the one you are reading edited professionally and corrected as I find money to do so . . .

In the book that deals with gravity there are just too many and numerously wide ranging facts that form the complete picture as a whole, which leaves me unable to include a full introduction in a space as small as that which page will allow. The explaining include for instance those phenomena, which I call the four cosmic pillars, but wise as you are, you would not believe me at this point that I have cracked the coconut because I guess in your vast experience you have seen too many idle explanations in the past proving to be senseless and little impressive, therefore my mentioning my success would not matter much either way. The proof I bring is true about gravity being formed as a result of these phenomena, **1) the Lagrangian system 2) the Roche limit 3) the Titius Bode law 4) the Coanda affect,** which I explain by delivering mathematical proof as to how they fit into the overall picture of gravity and which I mention just below. I prove the fact that every individual one of those phenomena is forming a unit that is in total being what we think of as gravity. The phenomena altogether constitutes a unit that forms the process working as gravity. Nevertheless my mentioning these facts will be just completely unbelievable to you without you reading the book, because I guess you have heard some attempt to explain the phenomena before but when I say you have not heard it in the context I put it, you might still be most sceptical because you have never heard it in the correct manner that I explain it and that poses the difference. Still you may not be convinced about my claims and although my explaining the phenomena is correct, does not change the fact that you don't believe me. The phenomena form an intergraded unit that results in gravity forming where each forms a part of gravity. You may still be you would be sceptical ...but convince yourself that I did manage to:

1) Find the location, position of singularity as a factor forming space-time
2) Finding space-time by dissecting Kepler's formula in relation to valuing singularity
3) Finding and proving space-time and aligning space-time with gravity
4) Find the working principals behind gravity as a cosmic occurrence.
5) Find the reason for the Roche limit and explaining the resulting of gravity from that.
6) Find out why the Lagrangian system, becomes the building form of the Universe.
7) Find why the Titius Bode law mathematically provides the foundation of gravity
By proving that the Coanda affect is gravity through activating space-time
By using the above the four cosmic pillars, it enable me to present the proof where I now can explain what conditions bring on the sound barrier. By proving it is gravity that the individual structure generates motion above and beyond the gravity the Earth provide is what is producing individual motion that the independent object earned within the sphere of motion that the Earth's gravity provides where the independent and individual motion put the relevance that gravity has beyond the conserving means gravity has where the space that is serving the independent object is independently in motion. The adding to the independence on top of the normal structural independence is creating more

individualism by the independent motion of the individual structure being apart from the motion that the gravity of the Earth provides. The fact every one misses is that any structure that is not part of the Earth's crust has an independent gravity and the form this gravity applies is stronger than the Earth's gravity which is why the structure maintains its form and this provides the independent individuality the structure has giving the unique structural space. The gravity of the Earth strives to incorporate everything into the Earth's sphere and into the Earth's structure and therefore the fact that the object is not incorporated into the Earth shows defiance and individuality, which gives it, mass.

By applying individual motion on top of the structural individuality that increases by the motion that the Earth provides, the independence of the individual object is becoming further exaggerated by having independent motion, which is further defying the incorporation the Earth strives to achieve. As the motion of the independent object grows more independent by applying more excessive motion to such an extent where motion creates almost the ultimate independence that may free the individual object with independence from the motion the Earth creates is what is breaking the restraint gravity has on all objects with independence formed by their structure. The structure show independence at all times by not forming part of the structure of the Earth within the sphere of the Earth's gravity. Moving about shows even more reluctance on the part of the top when spinning allows the top to eventually become part of the Earth. Breaking the sound barrier is the motion in space duplicating space by crossing over gravity borders, which is the limit to what constraint the Earth may produce in accordance with what full independence would allow.

These are the definitions underwriting cosmology and while my work is that much ignored; let's see how far I stray from these definitions in comparison of how much Mainstream science underwrites these definitions by bringing indisputable proof in presenting unwavering hardcore facts.
Quoted directly from the Oxford dictionary of Astronomy the following:

The definition of space-time is as follows:
Space-time is a four dimensional position of the Universe where the position of an object is specified by three coordinates in space and one position in time. According to the theory of special relativity there is no absolute time, which can be measured independently of the observer, so events that are simultaneous as seen from one observer occur at different times when seen from a different place. Time must therefore be measured in a relative manner as are positions in three-dimensional Euclidean space, and this is achieved through the concept of space-time. The trajectory of an object in space-time is called world line. General relativity relates to curvature of space-time to the positions and motions of particles of matter.

The definition of singularity is as follows:
Singularity: a mathematical point at which certain physical quantities reach infinite values for example, according to the general relativity the curvature of space-time becomes infinite in a black hole. In the big bang theory the Universe was born from singularity in which the density and temperature of matter were infinite.

The Oxford dictionary of Astronomy defines gravitation as follows
Gravitation is the force of attraction that operates between all bodies. The size of the attraction depends on the masses of the bodies and the distance between them; gravitational force diminishes by the square of the distance apart according to the inverse square law. Gravitation is the weakest of the four fundamental forces in nature. I. Newton formulated the laws of gravitational attraction and showed that a body behaves as though all its mass were concentrated at its centre of gravity. Hence the gravitational force acts along a joining of the centres of gravity of the two masses. In the general theory of relativity gravitation is interpreted as the distortion of space. Gravitational forces are significant between large masses such as stars planets and satellites, and it is this force, which is responsible for holding together the major components of the Universe. However on the atomic scale the gravitational force is about 10^{40} times weaker than the force of electromagnetic attraction

I have to give potential readers this fair warning that *The Cosmic Code as the Absolute Relevancy of Singularity* requires a somewhat higher level of understanding and needs a greater degree of insight that the other books in this series does namely

The Absolute Relevance of Singularity The Journal
The Absolute Relevance of Singularity The Unpublished Article
The Absolute Relevance of Singularity The Dissertation
The Absolute Relevance of Singularity in terms of Newton Book 0
The Absolute Relevance of Singularity in terms of Cosmic Physics Book 1
The Absolute Relevance of Singularity in terms of The Sound Barrier Book 2
The Absolute Relevance of Singularity in terms of The Four Cosmic Phenomena Book 3
The Absolute Relevance of Singularity in terms of The Cosmic Code Book 4
The Absolute Relevance of Singularity in terms of Life Book 5
The Absolute Relevance of Singularity in terms of Investigating Kepler Book 6
The Absolute Relevance of Singularity in terms of The Thesis Book 7
The Absolute Relevance of Singularity in terms of The Cosmic Creation Book 8

I have no chance that what I state as my theory on __*The Absolute Relevancy of Singularity*__ will be read, or much less that it will be seriously considered and I have not a snowballs hope in hell that it will be accepted by those with the authority to change physics principles. The theory I introduce here and now would never be accepted in my lifetime because science in the Newtonian way is bent on believing in the marvellous, the facts bordering the supernatural, the outrageously inconceivable and the magic of what can never be explained, although they claim to use facts. It is the marvellous to think that mass can create gravity. It is bordering the supernatural to think that with nothing between stars, yet by the magic of mass, mass has an unexplainable ability to attract another star many astronomical units away. It is the outrageously inconceivable to argue that life started on Mars, then overcame the quite impossible to escape the gravity that Mars holds on all things held captured in its gravity, and after overcoming the unthinkable, then made a dive for the Earth just to come and evolve over here. Science think they my have the ability to create a Black Hole in a Manmade atom-accelerator because science thinks of the Black Hole as the magic of what can never be explained and therefore that proves that science has no idea of what a Black Hole is while I can prove what a Black Hole is singularity reaching limits That fact that I can explain what a Black hole is, that ability the Wizards of Oz will never allow because the explaining I present is clashing with Newton head on and it is to be done in as simple manner as I am about to explain. However, when I prove what a Black Hole is I am going to destroy the fantasy world everyone makes believe as physics. To science a Black Hole is a world of magic where gravity has the ability to go mad and a Black Hole is something that man could manufacture by creating an atomic accelerator tunnel, or so science thinks. In other words the best science at present can do to explain the gravity in a Black hole is to give gravity a level of superior intellect and then take it away (by allowing gravity to go mad as it seemingly does in Super Novas and in Black Holes). Why can I prove what a Black Hole is…it is because I can prove what the Roche limit is and believe me that is one thing science this far could never get around in proving. The facts they use is as much fiction as Little Red Riding Hood's talking wolf…when it comes to explaining the integrating details of how gravity comes about. In science, when following my theory, everything can be explained by using physics, but using my explanation will make all present science become fiction, make all present science look like a fairy tale and make all present science seem to be good bedtime stories deprived of truth…and the money spent on Newtonian fiction-science will never allow me to have success because that would be too costly for the industry money-wise. Why would I call science a fairy tale…well this is just one of many, many reasons. Science wishes to promote something as impossible as time travel, which I show, is impossible. Science believes in travelling at speeds unlimited that could exceed the speed of light. I prove all such thoughts are impossible because I show that gravity and time is the very same thing. No one can beat gravity because gravity as time maintains the structural integrity of the Universe. In beating gravity one wishes to beat the cosmos that hold us secured. That is why time can manifest as what is known as the Hubble constant. Time is the redeploying of space by extending the absolute relevancy of singularity and that is only one of several factors that serve as time. Every time I declare Newton was mistaken and therefore science is wrong in presenting the most basics of physics, the workings of gravity, I am barraged by rejection and silent ridicule. Every time I challenge the Members of science to either prove Newton correct or to prove me wrong, I am ignored…my challenge goes unmet, so please forgive me for showing much antagonism…it is a result of Mainstream Science rejecting my efforts unfairly for many years. What I write is undeniably and undisputedly correct, but the instant science admits to my work being correct, that admission demotes most of the work science has accepted in the past as correct to the level of science fiction. It will destroy the groundwork of mainstream science and demote what is accepted to become fairy tales, which is what most Newtonian based theories are. Let Newtonian science explain what the cosmic purpose or the function is of a star…of a galactica…of an atom…of gravity…they have no idea. By the time you have finished this book you would have found answers to all the above questions in detail.

Mainstream science has so little idea of what a Black Hole is or what could cause a Black Hole that they devised a "Mini Black Hole" to suit there marvellous misinterpretations of gravity. That is a form of fantasy that fairy tale writers can't compete with. Science is so misguided in understanding life that they put life in all places throughout the Universe without ever finding one shred of evidence of the presence of life. Yet they say they work only with proven facts alone. They hold the opinion that life could have come from Mars but fail miserably in explaining how it will be possible for life to escape the gravity of Mars and then fly all the way, ever so precisely guided; directly to the Earth. How would it be possible for life to escape the gravity of Mars without them when explaining such a possibility by employing realistic physics, going into so much fantasy it leaves the story of the three pigs and the blowing wolf seem real. Science has the explaining of the exploding Super Nova down to the last detail where they explain that a Super Nova is gravity that has gone mad without ever proving how gravity can go mad because the truth of the matter is that gravity has no intellect to "go mad" in any way. Mainstream science always places new object found where their findings prove that the newly found object is on "the edge of the Universe", meaning where the Universe ends by forming an edge. This fantasy they dish up to anyone willing to believe him or her without ever telling what is beyond that edge. All they can see is an end of the Universe but in reality where there is an end there has to be a beginning of something else…this is physics. The Universe I show can't have an edge because I show where the point is that could never start and I show where the point is that could never end. I show that which can go no smaller and I show that which can go no bigger. I am about to introduce a Universe that mathematically can never start and the same Universe can mathematically never end.

I have been on a self-teaching mission that lasted thirty years and now that I have the answers and from which I have drawn the conclusions, I now find so much resistance from mainstream science in getting the findings my research uncovers out in the open. I offer tot academics many books in which I use diagrams, sketches, mathematical explanations and cosmic photos including other tools I employ to promote the required understanding needed to bring the ideas across that I wish to promote. However, publishing in this manner is very costly and money is one thing I do not have and therefore sending it to academics with no reply is an expense I cannot endure. Any academic feeling confronted by my accusation, please show how you prove

$$F \;=\; G\,\frac{M_1 M_2}{r^2}$$ is applicable and is true. Show how the use of the formula could be applied meaningfully to

present an answer worth of anything. Use the Newton's formula to show when the Moon is going to hit the Earth as the mass of the Earth pulls on the mass of the Moon. Better still, prove that mass does contract to create gravity and then explain how this is done…and please leave out the graviton because that is a joke! The idea

that mass draws mass closer $$F \;=\; G\,\frac{M_1 M_2}{r^2}$$ is mathematically proven as an untruth, which means it is not

true. What is the truth? …When you have completed this introduction you will have had a peeping view, a tiny glimpse of the truth…but as little as you would gain from reading this introduction alone, when put in comparison to what any person can gain from reading all of my work in total, you will gain endlessly more than what science is to explain about the truth, because what you then have gained by reading this document is much more than what science know about the truth. What I try to convey is that there is a good reason why academics block any and all publishing of my work, and when finishing this book, in comparison to what I offer, you have not even opened a first page of what I offer as new information when judging what my other work uncovers. Still, your effort in reading this document allows you to discover so much more of true science than what previously was known If you think I am boasting I challenge you to show where any of my explaining gravity requires superior intellect to understand... however in my simplistic approach to gravity I prove everything I say by applying the simplest mathematics there is.

The effort that this book represents the informing about an entire new way of cosmic appreciation meant to show that there are grounds for concern in the way science thinks and this book does not even bring all such arguments indicating concern in full. That one can only find when reading the first ten letters forming books named as with a title beginning with **Open Letters…**and those titles are included as books which I mention on my website, having the same name as this book namely **Amazon.com: Peet Schutte: Books** http://www.amazon.com/s?ie=UTF8&page=1&rh=n%3A283155%2Cp_27%3APeet%20Schutte.

I am about to prove that gravity is **the Coanda effect** and gravity comes about from four cosmic phenomena never yet understood since it was never yet explained. Science doesn't believe there is something such as **the Titius Bode law** but science does believe that mass would generate gravity. Science has no clue about **the Roche limit** but science believes in spite of the Roche limit that big craters on Earth are reminders of massive asteroids that hit the Earth in giant collisions. With the Roche limit in place these crates are the result of something else because it can't be from asteroids colliding with the Earth. We all know how the bicycle rides and we all think we understand how the bicycle rides but having the bicycle ride on two wheels have little to do with balance and everything to do with the Coanda effect.
The bicycle rides forward when peddled but also the bicycle rides downwards when peddled and the two are both linked to gravity. I am going to prove that the Coanda effect forms gravity. I am going to prove that the **Coanda effect** comes as a result of the **Titius Bode law**, **the Roche limit** and the **Lagrangian positioning system** but most of all how these are related to singularity. That means I am going to prove that mass has no effect on gravity but mass comes as a result of gravity. I am going to prove what singularity is and that there are two types of singularity that in the end is only one type of singularity.

Teaching ever since time began forms a pillar on which memory and remembering what you are taught is the most prevalent part of tutoring. One is expected to remember what those coming before and which are tutoring you, wish you to remember. The Tutor lays a foundation by ensuring that everything known and accepted coming from the past are well and truly founded in the mind of the student. In that there is no problem. The problem arises where the information studied is flawed and no one ever realised that. Fortunately this does not occur regularly, but if and when it does, notwithstanding the exceptional part it forms, it then becomes a major problem to deal with. Therefore what comes form the past are carried on into the future as unblemished truth and no person meddles with the thoughts called information given as study material. However, as unlikely as it could be, this did happen and it is part of the basis of physics. When the student is taught, the student is expected to accept without argument. What comes from the past are considered as tested beyond suspicion of inaccuracy and proven to what is absolute unwavering accuracy! It is way beyond doubt. There is this motto that students are mindless and students can only start to think after receiving information that came from the past.

Students are incapable of arguing by reason to introduce new thoughts. This ability to reason only comes after the learning process secured knowledge through the memory process and only when testing shows facts learned by memory is well established and it then forms a solid base for everything the student knows, then the students may form an ability to reason and to argue. This mostly takes about all the time that living one lifetime presents. Well, what happens when that everything that everybody believed in the present, inherited by all from the past, was totally flawed? This has happened to physics and no one in physics so far yet realised it. Not one in physics shows the ability to realise the flaw coming from the past as part of the legacy. Then the mistakes will carry on from forming facts the past, carried over as flaws into the future for as many generations as it takes to realise the mistake that is dragged along and this carrying on of a flaw could continue indefinitely, if there is no clear minds working to recognise the mistake and correct what needs to be corrected. I ask of you not to judge me according to what you have already achieved for in that sense I fall short of receiving your recognition in status. Judge what I present to you, for then you will realise with all my shortcomings, I present you with a truth that exposes short fallings in the basics of physics.

Here and now and before the beginning of what this document may be to any potential reader, all parties reading take note that I state it emphatically that all members forming the community of science in physics judges me being not sufficiently educated and certainly not to the level where I am able to form any opinion on matters concerning Sir Isaac Newton or his physics. Any and all of my self-tutoring goes begging in their eyes notwithstanding and regardless of the fact that I did my private and individual studies by which I furthered my insight. That allowed me to show with clarity what destructive force Sir Isaac Newton released in order to corrupt the laws of mathematics, contaminating science along the way and mostly raping the work of a great man, Johannes Kepler and what Sir Isaac Newton did to derail the truth and disguise scientific correctness where such violation can only be expressed as being blatant criminal fraud. What his deeds amount to, is to corrupt the laws of mathematics, to render the laws of cosmology useless and to rubbish all of science. By your reading, you will learn what it is that those academics that are guarding science never wanted published and read by the public at large. What I say is don't run and hide from my attack and coward away from my confrontation as so many of the most intellectuals amongst the Physics Paternity did when I confronted their thinking. On every occasion where I confronted members of the Academic Paternity in the past, those I confronted acted in precisely such a manner, such as cowardly ending all reading by throwing the book down, and then pretending to show the utmost disgust in what I say.

I researched the work of Kepler and found science doesn't even recognise his work, while it is his formula that forms the basis of all physics. Everyone thinks that Kepler found planets rotating, with Newton being able to explain Kepler, which makes everyone more concerned about how Newton saw Kepler's work. The formula used in physics as a principle is $F=mV^2$ which should be $F^3=mV^2$. $F^3=mV^2$ is replicating Kepler's formula in detail as $a^3=T^2k$. By using Kepler's formula we have $F^3=mV^2$ that is a precise replica of $a^3=T^2k$. The duplication is so obvious that we have (F^3 becoming a^3) while (m is k) and (V^2 is T^2). Einstein also only duplicated Kepler's formula by putting $E=mC^2$, which also should read $E^3=mC^2$. Again that is precisely Kepler's formula $a^3=T^2k$. (E^3 is a^3), (m is k) and (C^2 is T^2). In $E^3=mC^2$ Einstein mimicked $a^3=T^2k$, Kepler's formula. (E^3 is F^3 is a^3), (m is k) and (C^2 is V^2 is T^2). So what is so brilliant about Einstein's formula if Kepler had it centuries before? $E^3=mC^2$ is $F^3=mV^2$ which is $a^3=T^2k$. Newton corrupted the formula when he added $4\Pi^2$ to the formula and removed k that Kepler introduced while $a^3=T^2k$ Newton ignored. Newton changed $a^3=T^2k$ by using the symbols G ($m + m_p$) to replace k and then declared $a^3 = T^2$. I still wish to see the proof confirming Newton's changes as being correct notwithstanding that everyone thinks physics is entirely based on this conception. Whether the formula used is $F^3=mV^2$ or is $E^3=mC^2$, it still remains duplicating what Kepler introduced as $a^3=T^2k$. So I changed it back to Kepler's version of $a^3=T^2k$ as to better the understanding of the foundation of astrophysics and mainstream physics. The entirety of physics is not based on Newton. Physics precisely duplicates Kepler's findings while science doesn't even recognise Kepler's formula. By giving Kepler the credit due, the entire Universe becomes completely understandable...but then for my audacity to show mistakes in physics I am ignored flat! All I ever ask is prove the truthfulness of $G(Mxm)\div r^2$ because it is $F^3=mV^2$ that forms the basis of physics and that accuracy comes from Kepler's view of $a^3=T^2k$ that became Einstein's $E^3=mC^2$.

Whilst recognising the work of Johannes Kepler, Mainstream science bluntly ignores the impact of his work, and in that they miss the full vastness of the wide influence of his work. Newton shrouded Kepler's work under a blanket of alterations which I show was most unwanted since Kepler's work needs no alterations or corrections and every one since then kept Kepler's work hostage under Newton's changes. It is therefore almost absolutely realistic to say that all information what you are about to read in this letter and article sent to you for your attention was never yet printed in the near or the distant past although Kepler's work has been with us for about four hundred years, during which time it went unnoticed. It seems to me that any research predating Newton never came into use or into practise. My investigation of Kepler's work brought about a conclusion that no one yet arrived at concerning them with the findings of Kepler because no one scrutinised Kepler's formula before. Everyone is satisfied with Newton's version notwithstanding the incorrectness of it. The world seems satisfied with the idea that Kepler found planets rotating around a centre formed by the Sun and because of that Newton saw a circle. Where Newton saw a mathematical circle and was unable to understand $a^3 = T^2k$, Newton added what he thought is mathematically required to indicate such a circle. Newton added a mathematical $4\Pi^2$ to the formula of Kepler and removed the distance symbolising measure that Kepler introduced using k. On the other side Newton changed the symbol of k by using the symbols $G (m + m_p)$. This is just a longer and probably a more detailed manner of indicating k and better defining of k but it symbolises precisely to the point what k stands for nonetheless. I wish to draw your attention to the matter of Johannes Kepler's findings that Mainstream science considers as resolved and closed for many a century while it is not. My investigating Kepler helped me too resolve other unresolved matters but it was only possible by using Kepler's work. This brought about the idea that the Universe is in a state of contracting towards a centre of sorts where mass will form this contracting. This was prevailing until a man by the name of E. P. Hubble came to the forefront.

E. P. Hubble (1889-1953) confirmed an expansion through out the Universe, which contradicted all that science thought was known about our Universe. According to the accepted Newtonian cosmology everyone is of thought that the Universe is in a normal state of contraction because that is what $F = G \dfrac{M_1 M_2}{r^2}$ implies. Every person is very aware of the idea that the universal expansion would not last for ever, but has to start with some contracting effort at some point. Then all the heavenly bodies will collide and destruct, without any thought about any wavering on the matter and on the matters reliability there is evidently no doubt. When $F = G \dfrac{M_1 M_2}{r^2}$ apply, there should not be any force, which is able to keep the mass that is producing all the gravity that contains the Universe apart. Known for almost a century, science has failed to give any explanation about this cosmic phenomenon of a Universal expansion except for some silly notion about dark matter being dormant and not forming gravity as it should. If the dark matter is present as is claimed, then why doesn't the mass form gravity as it should and contract? What does our ability to see or not to see or the luminosity that the dark matter does not have, got to do with the mass bringing about pulling power, that is if mass brings about any pulling power. If the mass is there, visible or not, then the dark mass has to pull because light has no standing in the forming of gravity and if mass does pull, it has to pull to form gravity. However Hubble's law contradicts this idea of a collective contracting Universe totally. This phenomenon about Hubble's constant finding the cosmos expanding should not occur with Newton's perception about gravity envisaging the contraction that must come by the force created by mass in $F = G \dfrac{M_1 M_2}{r^2}$. If the Universe is on a contracting as Newton said it has to, we have to first find proof about the location to where such contraction is pointing. In order to locate the contracting we have to locate the centre of the Universe, which means we have to locate singularity. With singularity eternal small, holding the place where the Universe started, we first have to differentiate between singularity and zero, should we wish to find singularity. In modern science the phenomenon we know as the Roche lobe comes more and more to the foreground, indicating an undeniable interaction between orbiting structures sharing a common axis.

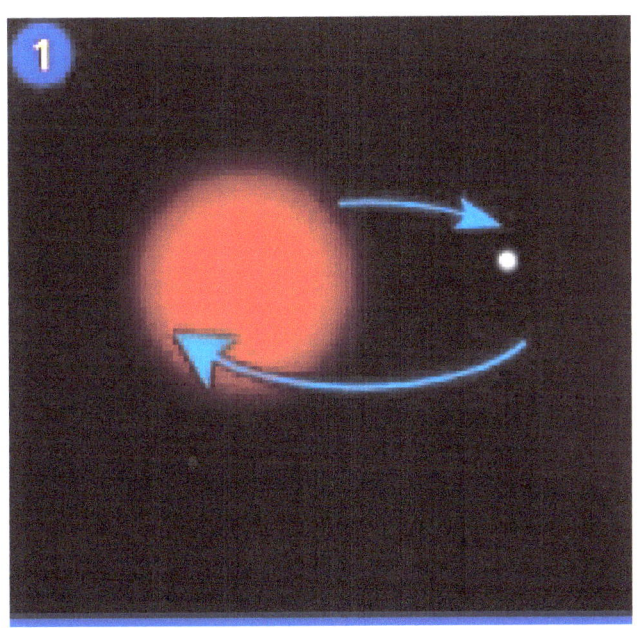 

That axis science at present does not recognise, notwithstanding the reality and undeniable proof there is behind all evidence. As apparent as it is to me, I went about divorcing $F = G\dfrac{M_1 M_2}{r^2}$ from all ideas forming cosmology and applying the roundness we have in Π to specific positions where one may locate singularity, which we have to locate if we wish to find gravity.

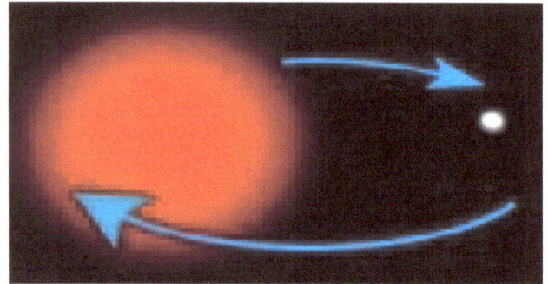

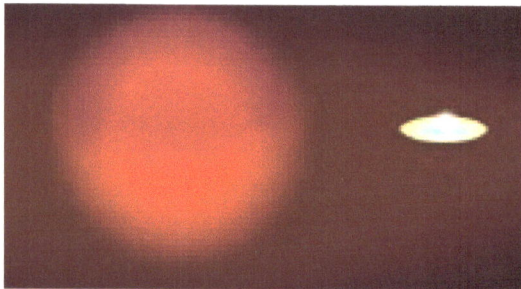

The Roche limit in the practical sense

The formula $F = G\dfrac{M_1 M_2}{r^2}$ cannot explain the comic occurrence shown in the pictures above called the Roche limit, I should find some attention when I say I can explain what is occurring in this instance and this occurrence connects directly to the Roche limit, as explained above. Not only does the Roche limit explain this phenomenon, but also it ties directly to the Titius Bode principle, also being another inexplicable factor in light of the formula $F = G\dfrac{M_1 M_2}{r^2}$.

According to the formula of $F = G\dfrac{M_1 M_2}{r^2}$ all orbiting structures should collide with a bang, but instead they do the tango until one drop, but when dropping it still does not collide with the larger structure, as would the formula $F = G\dfrac{M_1 M_2}{r^2}$ suggest that is used by science.

The position where the formula applies is most surprising. Where the formula $F = G\dfrac{M_1 M_2}{r^2}$ applies, one has to find singularity applying because the position of r is pointing to a specific pinpointing of space contracting.

The Coanda effect

The Coanda effect where a liquid concentrates around the surface of a solid and by movement concentrates the density of the liquid to gather and compact while maintaining a relevance to the centre of such a round solid. I discard the idea that mass could be responsible for forming gravity because in almost four hundred years all evidence is indicating that the truth is to the contrary.

This is not only limited to planets in our solar system. In the Universe, there are giant stars spinning around each other. These stars are binaries, which are also one form of double stars where double stars are another such a form. The difference between the types depends on the distance they remain apart. They keep a certain distance apart and do not collide. In the case of the sun and its planets, it could be a case that the systems might be to small, or they might be to apart. However, this is not the case with binary stars. They are close, they are big, and they spin around a mean axes called the Roche limit.

The Roche limit is:
The region surrounding each star in a binary system, within which any material is gravitationally bound to that particular star. The boundary of the Roche lobes is an equipotential surface, and the lobes touch at the inner Lagrangian point, L_1, through which mass transfer may occur if one of the components expands to fill its lobe. It names after the French mathematician Edouard Albert Roche (1820-83).

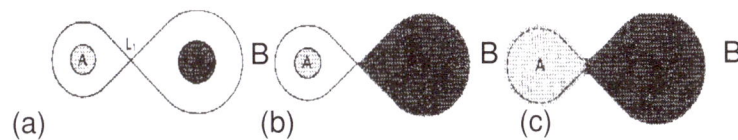

THE ROCHE LOBE: In a binary system, the Roche lobes of components A and B meet at the L_1 Lagrangian point. (a) In a detached system, neither star fills its Roche lobe. (b) In a semidetached system, one massive component, B, fills its Roche lobe. (c) In a contact binary, both components overfill their Roche lobes and share a common envelope.

LAGRANGIAN POINT:
The Lagrangian points
are five equilibrium points
in the orbit of one body
around another, such
as a planet around the Sun

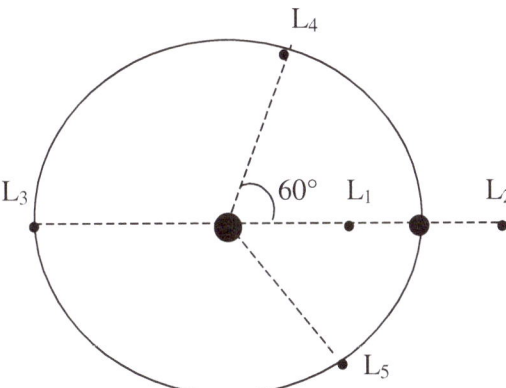

From singularity there comes three values each holding 180^0 and this fact science is familiar with. The straight line is always a potential triangle with on side apparent and the other side in infinity.

Planet	Mercury	Venus	Earth	Mars	Ceres	Jupiter	Saturn	Uranus
Bode's Law distance	4	7	10	16	28	52	100	196
Actual distance	3.9	7.2	10	15.2	28	52	95	192

Bode's Law:

A numerical sequence announced by J.E. Bode in 1772, which matches the distances from the Sun of the six planets then known. It is also known as the Titus-Bode law, as it was first pointed out by the German mathematician Johann Daniel Titius (1729-96) in 1766. It is formed from the sequence 0,3,6,12,24,48,96, and 192 by adding 4 to each number. The planets were seen to fit this sequence quite well – as did Uranus, discovered in 1781. However, Neptune and Pluto do not conform to the 'law'. Bode's Law stimulated the search for a planet orbiting between Mars and Jupiter that led to the discovery of the first asteroids. It is often said that the law has no theoretical basis, but it does show how orbital resonance can lead to commensurability. The importance that becomes known is the sequence the Titius – Bode law saw in the number arrangement of 3; 6; 12; 24; 48; 96 etc. The incorrect application of the Titus Bode law lies in subtracting the figure of 3 from 10 leaving 7. The other way of reasoning is to add four each time to the first value of three starting with 3 and so on. The true significance of the Titus-Bode law is that it points directly to a circular growth of 7 stages. The 7 relating to 10 is a precise derogative of the Roche limit or the Roche limit is a precise derogative of the Titius Bode principle because he two systems interlink.

The question immediately springing to mind is how on earth have I manage to come to find gravity being formed with the Titius Bode law applying. It is shockingly simple!

The importance that becomes known is the sequence the Titius – Bode law saw in the number arrangement of 3; 6; 12; 24; 48; 96 etc. The incorrect application of the Titus Bode law lies in subtracting the figure of 3 from 10 leaving 7. Then secondly is cross applying this into the number arangeemnt hels by the Earth in the Tiitius Bode law which is 7 and 10.

When looking at the Titius Bode law the alignment does not make sense because the distance doubles every time a new planet is positioned. Mercury is 3 and Venus is 6 and the Earth is 12 and in that the meaning of this is very much hidden.

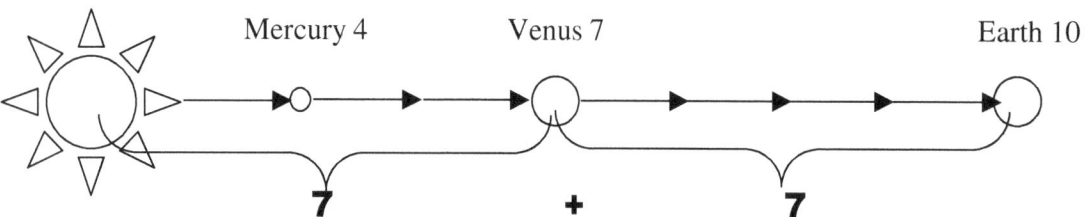

Looking at the Titius Bode principle and not the method we see that Venus, which is the Earth's immediate inner planet, holds a position of 7 in relation to the Sun and when this doubles we will find the Earth also holding a position of 7 from Venus, which the immediate inner planet is doubling from Venus to the Earth. If the distance doubles every time, then the frequency between Venus and the Earth must be the same as the distance or frequency between Venus and the Sun. In this same table the Earth holds a position of 10 in the method of measure applying. This puts the earth at a double 7 and also a factor of 10.

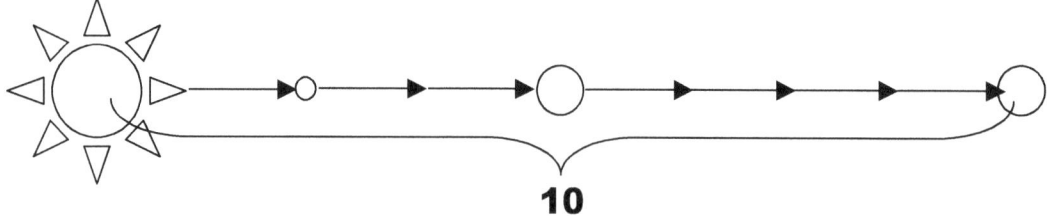

10

From this comes the value of gravity because as time used gravity, where gravity is time and gravity is used to form space, this forms the pattern whereby the building blocks were laid down by singularity to form space. The space we see is the remembrance of gravity applying that formed space as time formed gravity.

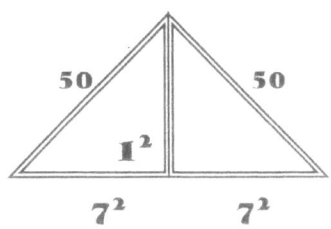
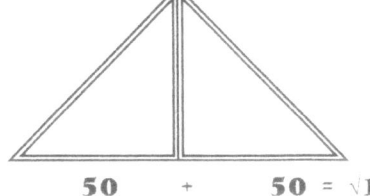
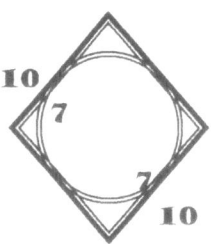

$$50 \quad + \quad 50 = \sqrt{100} = 10$$

Matter in relation (part of) to the total dimension of space.

$$\frac{\left(\dfrac{10}{7}\right)}{\left(\dfrac{7}{10}\right)} = 2.04 \qquad\qquad \frac{1.4285}{0.7} = 2.04$$

Taking from both orbiting influences

SPACE DIVIDED INTO TIME

$$\frac{\left(\dfrac{7}{10}\right)}{\left(\dfrac{10}{7}\right)} = 0.49 \qquad\qquad \frac{0.7}{1.4285} = 0.49 \qquad \boxed{\text{Taken from both orbiting influences}}$$

SPACE MULTIPLIED WITH TIME

$$\frac{\left(\dfrac{7}{10}\right)}{\left(\dfrac{10}{7}\right)} \quad \times \quad \frac{\left(\dfrac{10}{7}\right)}{\left(\dfrac{7}{10}\right)} = 1.$$

Therefore no influencing and relevancies doesn't change.

THE PROCESS PARTED USING THE ROCHE PRINCIPLE

$$\frac{\left(\dfrac{10}{7}\right)}{\left(\dfrac{7}{10}\right)} \text{ Motion of divisions counterbalancing}$$

$$\left(\frac{\Pi}{2}\right)^2 \text{ The Roche influence on 2.04 Titius Bode}$$

$$2.04 \times \left(\frac{\Pi}{2}\right)^2 = 5.033$$

$$\left(\frac{\Pi}{2}\right)^2 \times 2\left(\frac{10}{7}\right)$$

Bringing into the equation the Roche effect

$$2.04 \times \left(\frac{\Pi}{2}\right)^2 + 2.04 \times \left(\frac{\Pi}{2}\right)^2$$

$$= \ 5.03 \ + \ 5.03 = 10.06 \text{ from both objects}$$

In case of the solid we have Π^0 by seven and in that we find that Π^0 is substituted by the number of $7\Pi^0$. When we find 7 stands related in terms of $\frac{7}{10}$, which then holds the value of $\frac{7}{10}$ (space-time on one side of the divide) the value of space-time concerning the solid part of Creation is then $\frac{7}{10}$. That means on the solid side we find that Π is replaced by the Titius Bode value applying as $\frac{7}{10}$ and in the square it is the square of material turning inside liquid which is $\frac{7}{10} = (0.7)^2 = 0.49$ on both sides of the divide $\left(\frac{7}{10}\right)^2 + \left(\frac{7}{10}\right)^2$.

As material spins within the liquid, we then have the solid moving from $\frac{7}{10}$ to a new position of $\frac{7}{10}$, making $\frac{7}{10}$ the value of Π and the gravity value of $\frac{7}{10}$ becomes the square of $\frac{7}{10}$, which is (0.49). This happens to the top of the sphere as well as the bottom of the sphere and combining to movement altogether gives a motion value of $0.49 \times 2 = .98$

= 0.49 + 0.49 on both sides of the divide.

= 0.49   on both sides of the rotation

= 0.49 × 2 = .98

= 1   also on both sides of the divide.

= .49   also on both sides of the divide.

.49 + .49   =   .98

$.98 \times 10.06 = 9.86 = \Pi^2$

$9.8 = \Pi^2$ Motion or gravity

Motion or gravity or in truth TIME

= SPACE - TIME = Π^2 = 9.86

The ratio of 7 to 10 would apply as seen from every planet as the planet circles the Sun. The fact that we see 7 to 10 applying is because we are within the governing singularity of the Earth by forming a part of the controlling singularity of the Earth. The same ratio of 7/10 will apply when standing on another planet.

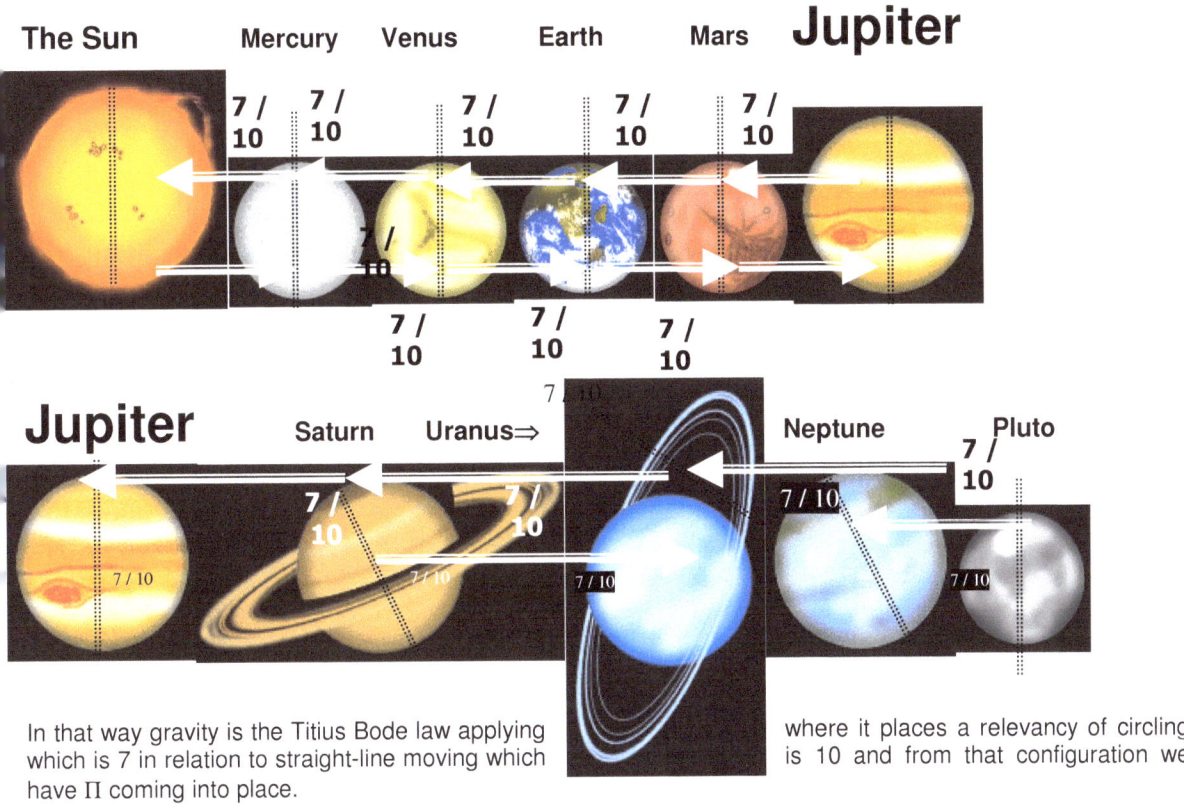

In that way gravity is the Titius Bode law applying which is 7 in relation to straight-line moving which have Π coming into place.

where it places a relevancy of circling is 10 and from that configuration we

Inclining by 7°

Sun

Earth

Inclining by 7° as the Earth goes around its axis.

I repeat this very basic law because it is evident that Professors forget the most basic mathematical principles such as that multiplying zero with anything leaves only nothing therefore zero is not a mathematically usable number; still they want to fill the Universe with 0. In mathematics gravity is the **Pythagoras's theorem**, which is a relation in Euclidean geometry among the three sides of a right triangle. I am sure most know this but for those professors that forgot how this works since they forgot so much of the most basics of mathematics, this is how it reads: **The Pythagorean theorem**: The sum of the areas of the two squares on the legs (*a* and *b*) equals the area of the square on the hypotenuse (*c*).

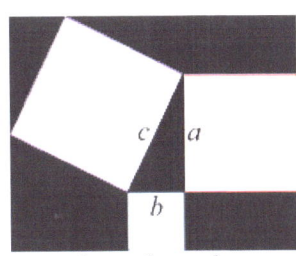

The theorem is as follows: In any right triangle, the area of the square whose side is the hypotenuse (the side opposite the right angle) is equal to the sum of the areas of the squares whose sides are the two legs (the two sides that meet at a right angle). This is usually summarized as follows: **The square of the hypotenuse of a right triangle is equal to the sum of the squares on the other two sides. If we let c be the length of the hypotenuse and a and b be the lengths of the other two sides, the theorem can be expressed as the equation:**

$a^2 + b^2 = c^2$ or, solved for c: $c = \sqrt{a^2 + b^2}$.

This equation provides a simple relation among the three sides of a right triangle so that if the lengths of any two sides are known, the length of the third side can be found. A generalization of this theorem is the law of cosines, which allows the computation of the length of the third side of any triangle, given the lengths of two sides and the size of the angle between them. If the angle between the sides is a right angle it reduces to the Pythagorean theorem. This says we can solve the riddle of the Titius Bode law!

Everything about singularity is in a long line that is equal to a triangle that is equal to a half circle, and when something turns it is not a circle and an axis but it is seven positions that changes.

$$\Pi = \frac{7+7+7+0.991}{7}$$

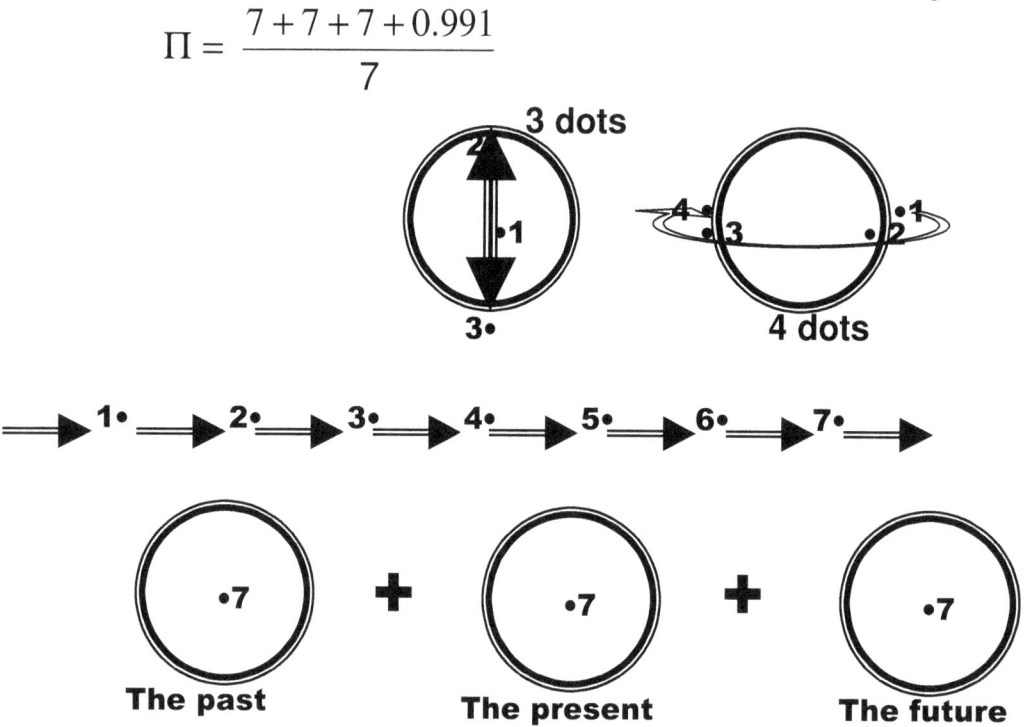

We have the Earth spinning by 7° around singularity, once spinning around its axis ands the spinning around the axis of the Sun. The 7° are movement and therefore by being movement it has to be calculated by the square thereof. Spinning around singularity brings equality to 7 duplicating because 1° = 1°.

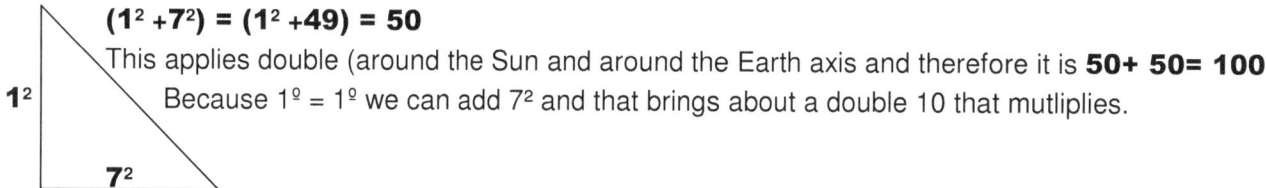

$(1^2 + 7^2) = (1^2 + 49) = 50$

This applies double (around the Sun and around the Earth axis and therefore it is **50+ 50= 100**

Because 1° = 1° we can add 7^2 and that brings about a double 10 that mutliplies.

When we take the square root of 100 it becomes 10 and there the Titius Bode law applies as the law of gravity and as the law of time moving by forming space forever. The Titius Bode law shows the Earth is circling by 7 and is moving forward by 10 and that forms gravity that forms space. From this we can deduct that the Universe moves 7/10 in a three-dimensional sphere (Π^6) forms as it starts at $7/10(\Pi^6)\div6 = 112$, which is a value forming the start of the element table and that I explain in the Cosmic Code. One is 7/10 which is the Titius Bode law which is the interaction of gravity spinning and by spinning is forming a sphere (Π^6) within a cube ($\div6$) and that is how the cosmos forms using Π. The dimension of $\Pi^0\Pi$ is flat but by spinning $\Pi^3=(\Pi\Pi^2)$ using 6 dimensions, the Universe goes in a sphere (Π^6) spinning in a cube 6.

In this I prove that for instance amongst so many other things that the Universe is a sphere spinning in a cube. By ticking $\Pi^0\Pi$ time forms space by becoming space as time moves into the future leaving the past behind as space. This is what we see from how the Titius Bode law is employed whereby the Titius Bode law is the way of building the solar system just as it then has to be the building of the cosmos. Time is a substance and is the only renewable substance with the ability to come into the Universe because from the start it came into the Universe to form the Universe as space.

Time renews what is by securing what was as space. As time moves on space grows by the margin of singularity $\Pi^0\Pi$ leaving spots that form dots. The proof of this is in the value of Π being 3.14159 where 3.14159 -3 = 0.14159 x 7 = 0.9911, which is singularity as the spot (0.9911) becoming singularity 1 as the dot.

That is gravity and not some idiotic cry about mass performing magic to bring on gravity.

Now every physicist, show your academic worth and your educated dignity and accept the challenge I put to you and to all physics educators: I challenge you all: **PROVE ME INCORRECT IN ANYTHING I SAY!**

Whatever gravity is, gravity has to be Π. If gravity is linked to mass as Newton stated, then mass has to be very closely connected to Π. Looking at every aspect that forms gravity, it is formed by a circle. The Earth as much as the Sun as much as all stars and galactica holding gravity is round and the roundness are Π. The curvature of space-time, the fact that gravity bends light into a curve, this bending comes in the form of a circle that is formed by Π. The Sun for instance spins around and that is formed by Π. The Earth holds the Moon captured while the Moon circles around the Earth and the circle is a result of Π. If it is with gravity that the Moon circles around the Earth, then in all of this we must locate gravity holding Π as a value.

Looking at the Solar system we find that all planets and objects not classified as planets and all things that is just simply forming solar debris has one thing in common…all apply the value of Π in the process where they orbit the Sun, which also uses the formation value of Π to construct the roundness the Sun has. Gravity has much more in common with Π than it will ever have with mass that produces gravity. Wherever singularity forms gravity, it involves Π which then results in gravity manifesting as some or other form holding Π as a major factor.

Being at   Π **Going to**

Singularity $= \Pi^0$

Coming from $= \Pi$

Singularity $= \Pi^0$

Singularity: a mathematical point at which certain physical quantities reach infinite values for example, according to the general relativity the curvature of space-time becomes infinite in a black hole.

Where singularity holds position in the centre of any and all rotating objects as a value of Π merely applying movement (in the form of atoms) qualifies all matter to be space-time. It does not only fit the description of space within Black Holes but it fits all stars where singularity becomes part of all the stars from the minute to the largest cluster of matter.

With no line starting from zero because there is no zero as a mathematical fact, then all particles hold the point of infinity and not merely the Black Hole,

Through rotation encircling the point of singularity and matter is (1) coming from, (2) being at, (3) as it is going too in one movement in relation to the specifics of the centre point being singularity, all matter then qualifies to form space-time.

From that argument one may conclude that all stars will become Black holes depending on the gravity increase they may generate.

Stream of water

Stream of water

The Coanda effect #2
JL Naudin - 09-26-99

The Coanda effect #2
JL Naudin - 09-26-99

I say this phenomenon called the Coanda effect is gravity. I say mass is a product of gravity whereas Mainstream Science has been saying for centuries that gravity is a product of mass. Science says that gravity is due to mass establishing gravity while not one person could ever explain the least detail as to how it is done. I went on to research Kepler and I discovered gravity through discovering Kepler. I concluded that gravity is the movement of material through space. By following Kepler's guide as Kepler formulated the process in introducing the equation four centuries ago being $a^3 = T^2k$ he gave us an explanation to what gravity is…if only Newton took notice of this important document. This says material holding space moves through space and proves that gravity has nothing to do with mass while mass is the product of space moving.

What is it about gravity that I say which no one wants to know? No one wants to listen to my point because I call Newton a fraud. He defrauded science and took all the other suckers running after him like sheep that are / were unable to think by there own ability. Now no one wants to find out how stupid the entire lot was that came after Newton and followed in his misguided footsteps. Saying this much in the past had every academic rejecting my work at that point. N academic found my work worthwhile to read after reading this much about Newton. No one wants to know that Newton went on lying for almost four hundred years. If you feel annoyed with my remarks concerning Newton, then explain how mass brings about gravity! No one understands the issues of mass and gravity. No one in science clearly distinguishes between gravity and mass and everyone in science tries to confuse the two issues by making them one and the same. They are two distinct different issues never to be confused.

I have discovered that the Universe is not employing a Special Relevance of singularity, but there is a state of **_The Absolute Relevancy of Singularity_** that is not only controlling the Universe but is what the Universe constitutes of...it forms the Universe ...it is the Universe. However, notwithstanding the magnitude in significance **_The Absolute Relevancy of Singularity_** poses to science forming a breakthrough, yet past experience taught me I have no chance that my theory on **_The Absolute Relevancy of Singularity_** will be noticed. I came to the conclusion that members forming the body of Mainstream science in physics will not care to take any notice of **_The Absolute Relevancy of Singularity_** and I don't believe that it will be read, will be seriously considered and much less be accepted by those with the authority to change physics principles. I hold the opinion that the theory I introduce here and now would never be accepted in my lifetime because science in the Newtonian way is bent on believing in the marvellous, the outrageous and the magic of what can never be explained, although they claim to use facts as a basis. Science has no idea of what a Black Hole is and I can prove what a Black Hole is. Science has no idea what "the sound barrier" is and I can prove what it is. The explaining of science coming from this that I prove is almost endless.

Yet, I feel I need to warn you whom are reading this letter that this work contained in this letter strays widely from mainstream science and for that there is a very good reason, but I should add that in the least it is thought provoking. I researched the work of a man that is most exceptional and even more prominent in the history of mankind and yet the meaning of his work went unnoticed all this time. His role in the gathering of information furthering knowledge accumulating of the human species' efforts stands second to none while most of everyone is not even aware of the full implication of his work. While recognising his work Mainstream science bluntly ignores his work and in that they miss the full vastness of the wide influencing of his work. It seems to me that any research predating Newton never came into use or in practise. My investigation of Kepler's work brought about a conclusion that no one yet arrived at concerning the findings of Kepler because no one scrutinised Kepler's formula. Kepler found planets rotating around a centre but Newton saw a circle and added what is mathematically required to indicate such a circle. Newton added a mathematical $4\Pi^2$ to the formula of Kepler and removed the distance symbolising measure that Kepler introduced using **k**. On the other side Newton changed the symbol of **k** by using the symbols G (m + m$_p$). All of this I change and show why it has to change back to Kepler's vision in order to better man's insight into physics, but in that I change the grain and foundation of mainstream physics, I change the total understanding of what forms the basis of cosmology and that part is what mainstream science avoids.

Not withstanding this, still I hope that this writing may spark interest even at such a low academic level and grade in scientific sophistication development because I am about to prove that I discovered:
1) The location, the position and the value of **singularity** as a factor forming space-time
2) Finding **space-time** by dissecting Kepler's formula in relation to valuing singularity
3) Finding space-time, **proving space-time** and **aligning space-time** with gravity
4) The **working principals** behind and manifesting **of gravity** as a cosmic occurrence.
5) The **Roche limit** and explaining the resulting of a law coming about from singularity.
6) The **Lagrangian system**, how and why that becomes the building form of the Universe.
7) The **Titius Bode law** and I show mathematically how gravity comes about from that
8) The **Coanda effect** and the producing of gravity through reproducing space-time.
9) The **sound barrier** by proving it **is gravity** generated **by motion** in space becoming independent where motion creates independence. Breaking the sound barrier is the motion in space duplicating space by crossing over gravity borders. It is $a^3 = kT^2$ where $(k \leq T^2)$ or $(k > T^2)$. Most of all, I prove that gravity is the Coanda effect forming, applying as gravity everywhere in the cosmos.

Kepler said $a^3 = T^2k$ but that could also be $k = a^3/T^2$ and could be $k^{-1} = T^2/a^3$ and that is the Coanda effect.

Newton said a sphere is $a^3 = 4/3 \, \Pi \, r^3$, which is mathematically correct, however

Kepler said the cosmos told him a cosmic sphere is $a^3 = k \, T^2$ where that puts the cosmos in completely different mathematical dynamics altogether. There are the two distinct possibilities of **a^3** which Newton saw and which Kepler saw and both are most valid, but altogether unequal. Between the two concepts there is literally one Universal difference and the two can never be mistaken as promoting the same principles.

It is true that Newton's method or formula of calculation $a^3 = 4/3 \, \Pi \, r^3$ when measuring the sphere is widely used, but Kepler received his code of calculation from a very high authority, which is none other than the Universe and therefore can not be discarded as Newton did. It is the duty of the cosmologist not to reject Kepler's findings, or as Newton did, try to transform it into something that Newton could understand after such transforming, because it then strays from the original meaning…but dutifully to search for the meaning as Kepler received the formula from the cosmos. We can test any of the following symbolic values in the mathematical expression and also test the principal behind the expression in which Kepler stated them. By such testing we will find that time after time there were never any corrections in the translations of Kepler's formula required since the translation thereof was never incorrectly presented by Kepler in the first place and in that a case therefore asked for no alterations as Newton did as to secure the correct reporting of the cosmic information being translated. By taking the formula on face value it can change as follows: **$a^3 = T^2 \, k$** can become **$k = a^3 / T^2$** or become **$k^{-1} = T^2/a^3$.**

When translating Kepler's mathematical expression into English we can see what Kepler said also read as **$k = a^3 / T^2$** where **k** is indicating one point from a centre point that is space **a^3** relating to time **T^2**. From a centre comes space-time. The centre **k** brings space **a^3** in ratio to time **T^2**, which are space / time **a^3 / T^2 k.** Reading this correctly cannot bring any dispute…yet it does…and it's been doing it for centuries on end!

Kepler was the very first person to mathematically introduce **space a^3** aligning a **centre k** and relating the resulting movement to **time T^2**. Not only did he introduce **space-time $a^3 / T^2 k$** but he also placed **space a^3** and **time T^2** in a relevancy **k** long before Einstein did and placed **gravity in space-time $a^3 / T^2 k$** even before Newton named gravity. He showed that space **k** is growing in relevance (**$k = a^3/T^2$** and also the opposite as **$k^{-1} = T^2/a^3$**). The manner in which the Universe attends to space-time is **$a^3/T^2 = k^1$** and **$T^2/a^3 = k^{-1}$**. Kepler was the person who placed gravity as the ingredient in the Universe that determines **space a^3** and **time $T^2 k$** and this proves the Universe is not expanding which it can't do, but is changing relevance by allowing material to grow as time develops space occupied. Kepler was the first one that saw that gravity comprises of two factors being **k** or linear gravity and **circular gravity or T^2** as gravity keeps space in form while all is staying together.

Kepler said **$a^3 = T^2 k$** and that correctly translates to a mathematical expression **$k^0 = a^3 / T^2 k$** which in the verbal statement in English translates that Kepler said that there is a **space a^3** which is **equal =** to the motion in **the time duration T^2** thereof between two specific points which holds a relation onto a centre **k^0** where from there forms **a straight line k** that is centred on the spot where space begins from **k^0 that produces k** as well as producing the circle therefore that spot **$k^0 = a^3 / T^2 k$** has hold **k^0** at a value of having the least space. The line **k** is centred onto a spot where space begins specifically at **k^0**. This point not only produces the line **k** coming from a point **k^0** but represents also the space **a^3** that forms the eventual circle by the rotation of **T^2**. Therefore from the centre holding **k^0**, **k^0** leads to **k** that forms the revolving space **a^3**, which is rotating **T^2** at a distance **k** where **T^2** forms the outer limit of **k^0**. Mathematically **$a^3 = T^2 k$** will also be **$k^0 = a^3 / (T^2 k)$** because **$k^0 = 1$**. But **$k^0 = 1$** also present the single dimension where all factors are a product of one. If anyone can locate **k^0** then also that person will find singularity. That is where gravity is because gravity is strongest where space is least. Then that suggests that gravity is strongest at **k^0** because there space is least. That is gravity because that is what keeps the orbiting objects in orbit but also that is what Newton completely missed when he changed Kepler's work. Newton failed to recognise gravity as the only ingredient in Kepler's formula. He admitted he missed this because he admitted he did not know what gravity is while Kepler explicitly showed what gravity is. Gravity is what keeps the orbiting objects in rotation while orbiting. **$k = a^3 / T^2$** is **distance^1 = space^3/ time^2** forming from a pivoting centre **k^0**. That is a cycle and moreover it is a cycle formed **by space/time**. What Kepler said is that space is **a^3 being in motion T^2 k.**

That says **space^3 (a^3)** relates directly to **time^2** that uses the symbols **T^2 k**. This is also what I refer to when I say one has to read what Kepler did not say when one wishes to see what Kepler meant to say. Kepler introduced space³ –time²⁺¹ long before Einstein's date of birth appeared on any calendar although Einstein is credited with the formulating of the concept of space-time and giving it a name. Going even further Kepler stated that the space **a^3** is on the move **T^2** around in a circle at a distance **k**. That is what that comet is doing. The space³ (Comet) is circling **T^2** the Sun using a radius **k** to establish the cyclic time² as a period of continuous motion and continuous motion is gravity. Remember in this statement I am separating cosmic principles applying from the way that gravitational principles apply on Earth.

As Kepler said **$a^3 = k \, T^2$ and therefore $k^0 = a^3 / k \, T^2$** and therefore we have to find **k^0**. As a result of examining this proposition, I located two principle positions both holding singularity. ___What is in the Universe is spinning___. **The entirety of everything forming the Universe is spinning inside the Universe and** such spinning are always in the centre of one specific point, wherever such a point might be. In the **precise middle** of all **objects in rotation** is a precise centre where this pre-designated centre is dividing the object in rotation into sectors that will **start the spinning initiation** from that centre point. By spinning the one side is coming towards while the opposing side at that time is going away. Thus, the spinning object **will have a middle point**, a very specific **centre point that does not spin** and only holds Π as a specific value because within that centre being that small, no radius can apply. But also within the one value forming, such a line **cannot have is zero** because the line **is there and holds contact** to the rest of the material bringing about that **zero does not start any** line and therefore the **value of the line must be infinite**, just as described in **accordance** and by **the definition of singularity.**

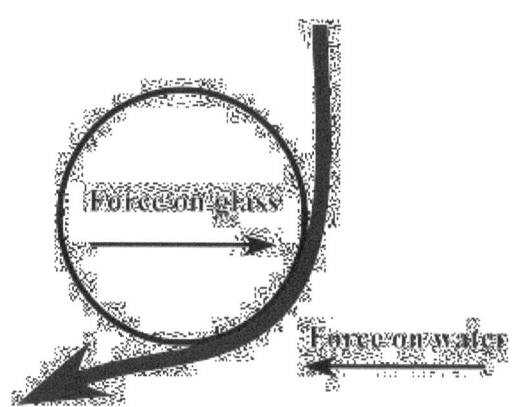

The condition for the presence of this centralised singularity $k^0 = a^3 / (T^2 k)$ **is movement** $T^2 = a^3 / k$ **of space** $a^3 = k T^2$ **in relevancy** $k = a^3 / T^2$ **going both ways** $k^{-1} = T^2 / a^3$
(Newton's 3rd law.)

This explains the Coanda effect and the Coanda effect is gravity and gravity "glues" the water to the glass!
In considering the spinning motion in the fraction of time in the detailed instant every aspect of rotation will turn in every instant of change in time by putting every spot there is in another location in accordance with the centre point that is unable to spin. While spinning the points will change direction every 90° of spinning and will oppose what it was every 180°. Although the points had the same characteristics only one instant before, they oppose the characteristics it had just before and just after the very instant in which they are and to which they relate by similar points also in rotation. The fact of the graph proves my point in quarterly opposing dimensions and values. As every point relocates, therefore every point completely changes its attitude from what it was to what it is in terms of what it will be when it is going there. Going down to the centre, as the rotating direction moves inwards, the rings will become smaller and smaller. In dimensional terms, which I explain later on the value of **2k** relates to **T^2**. That relation extends to the next value where **T^2** relates to **k**, which relates to **T^2**. The first space in the circle will then be **T^2k**. From the centre being in infinity one can realise by applying mental power the single dimension factor not seen but present all the same. Extending that into the 3D comes six **k** and any one of the six will further extend to form a seventh point as **T^2** All this is a multiplying of $k^0 = a^3 / (T^2 k) = 7$

Lines mathematically cannot start at zero because there is no evidence of zero as a factor in mathematics. Should you disagree with my statement the question in need of answering is this: **What will the length of the shortest hypothetical line imaginable be and moreover, what would the total overall length be in that case?**

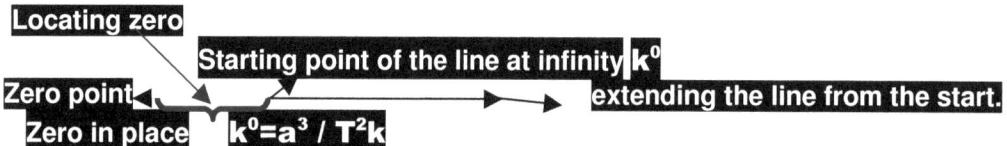

Let's find $k^0 = a^3 / (T^2 k)$ and see where it is hidden. The sphere is a circle in many facets and therefore we will approach the sphere as one multi dimensional circle, however the sphere as such remains one circle to the power of many. When investigating a circle one would draw a line from one edge running through a centre all the way to the other edge. In doing that we would find the measure of the diameter, which is most important when trying to establish the volumetric worth of the sphere. The circle has Π to indicate form and uses r² to establish the worth of such a circle by using the radius symbolised as r in drawing a straight line.
In any circle or sphere the size only depend on the fluctuation of r in the square as a component to the circle or sphere but that does not affect the form which comes by indication of Π in any way there may be. The conclusion from this is that no line can start at zero because that will be a mathematical impossibility.

This statement by itself excludes zero and with zero excluded one then begin to appreciate all the rest of the concepts governing corrected cosmology. A line or spot starting at zero would therefore be shorter than the shortest line possible. For obvious reasons can no line, or any line grow or extend from zero because such a line must then quit zero and become something, thus abandon its original value by the adding of the first value. Mathematically said it would be as follows 0+0=0 whereas of it started with something infinitively small it would be $1^0 + 1^0 = 2$ and then from using something infinitively small it will grow into something immense such as the Universe.

In any circle or sphere the size only depend on the fluctuation of r in the square as a component to the circle or sphere but that does not affect the form by indication of Π in any way there may be. The conclusion from this is that no line can start at zero because that will be a mathematical impossibility. If a line started with zero, that would nullify Π ($0^2 \times \Pi = 0$) and that would leave the form without having any form because $\Pi \times 0 = 0$. This statement by itself excludes zero and with zero excluded one then begin to appreciate all the rest of the concepts governing corrected cosmology. A line or spot starting at zero would therefore be shorter than the shortest line possible. For obvious reasons can no line, or any line grow or extend from zero because such a line must then quit zero and become something, thus abandon its original value, should such a line wish to progressively become more of what it was before.

If the reader is wondering where this is going, well I am trying to remove zero from lines as I am trying to remove zero from graphs as I am trying to remove zero from the Universe because I am trying to remove zero from filling outer space. When a line starts off with zero while still forming a line that composes of a line, that would mean the start of the line has a different value to the end and a line holds conformity through out…no line can end by applying zero as the concluding worth. When any line is starting from point zero it can never leave zero because of the influence of being zero disqualifies any possibility of growth, or even being present at any point to grow. If the line then had to grow in all directions at the same pace the line must therefore be a circle or being three-dimensional, a sphere. Flowing from this fact is that in the Universe there can be no zero point or unfilled space as every point has to be filled with at least something other than zero. In the case of the growing sphere the value of the circle is Π, and that has to be the point where creation started. That gave me the clue where to start looking for singularity. One would find singularity in the value Π and the value Π will be in all things rotating in a circle. You might wonder how does that apply to the cosmos and moreover to gravity? Mainstream science promotes the idea that outer space as far as outer space goes, is filled and even overflowing with nothing. The nothing they place in outer space is so much of nothing that the nothing they have filling outer space is overflowing because the nothing is expanding it is growing! That nothing then must be growing because E.P. Hubble proved the cosmos is expanding. If this sounds ridiculous then might I remind everyone that I didn't fill the Universe with nothing because that was the doing of mainstream science? You cannot fit nothing into outer space because it just will not fit; there is just too much space to harbour nothing. If any of the factors in Kepler's formulae represent nothing then that is what you will get, it would be nothing.

The Universe in its total entirety will be nothing. **$a^3 = 0$ $T^2 = 0$ k=0, which then mathematically could only translate to $a^3 \div T^2 k = 0$.** Or in the case of Newton removing **k** then Kepler's formula would read as follows **$a^3 \div T^2 k = 0$** because **$a^3 \div T^2 \times 0 = 0$.** If the argument seems ridiculous it is not my mentioning such a fact that is ridiculous but the mere fact of the reasoning that also became an accepting of the valid ness of recognising that it is nothing that fills the cosmos that is the silly part. The basis of such an argument and the fact it could be accepted by science is what is making it ridiculous. It is the fact that one must argue about such a ridiculous matter about an idea that nothing is overflowing while it is filling the Universe, that allows the ridiculous part to enter the conversation because the trend reminds of arguing about fairies and little people existing or not and such argument is nonsense. If space is nothing then explain how it is possible for nothing to have a distance and to have a measurement indicating the number per measured unit in length. Add as many zeros as possible and see how that can form 149×10^6 that the Earth is from the Sun. Using zero the distance indicated would be indicating just that value being zero or the capitol O indicating zero while every planet has a precise distance it is located in terms of the Sun as well as in relation to each other.

Try and indicate what is measured and calculated in value in outer space in measured space, while having that going in kilometres or astronomical units and then finding it is nothing in multitude filling that distance. The distance between the Sun and Pluto **is 5900×10^6** kilometres of space, but in that statement we take it that the one as a factor used in determining whatever constitutes to form the measure of a kilometre. By adding one **5900×10^6** times puts 1 present in such a multiplication because adding 0 **5900×10^6** times will still amount to 0. The one constitutes the presence of a fact being a statement of a value compiling to present the measure of space as it is in a distance. By saying the distance constitutes of nothing we have to substitute the one factor with a factor of zero. Then the calculation must read **Pluto is $5900 \times 10^6 \times 0 = 0$.** Including nothing as to state the presence of that part contained by the calculation delivers the total of zero. It seems as if science has ignored this issue by simply not thinking about the fact and therefore simply ignoring that what is measured forming the sole value of space has a practical worth, but it is somehow more convenient to put the value of nothing as part of the distance in calculation because that is what is measured and then see how one can multiply by using zero in mathematics and reach a distance holding a value other than zero when multiplying by zero.

I agree that what is filling outer space is invisible, but also it is there, it is present and being present and there while being invisible disqualifies whatever is there from being zero because being zero will mean it is not there and we cannot deny whatever is there of being there. Then what is there will be there, while being invisibly small, but it will still be possible to form a line because every aspect of the Universe forms lines while also it will have the potential to fill space and can still form a measurable unit. That then must be 1 because while $1 \times 1 = 1$, $1 + 1 = 2$ and that qualifies that invisible thing to be present ($1 + 1 = 2$) but at the same time be completely invisible ($1^3 = 1$). When realising this I knew what has to be true about that which I was looking for and that it had to be singularity because singularity can only have one value and that is 1.

To find the invisible I had to locate singularity. I realised that my effort to locate the point holding singularity enabled me to backtrack the exploding Universe to its origins. The Universe is a sphere because it is filled with spheres filling the void spaces (not the nothings) and in that I first had to investigate the visible.

Newton's mathematics says a sphere is $a^3 = 4/3\Pi r^3$ while Kepler said a sphere is $a^3 = T^2 k$, and both are equally correct because the cosmos gave numbers to support its statement.

With Kepler $a^3 = T^2 k$ and with mathematics the volumetric size of space must either be according to the measure of normal mathematics if it is a cube then three sides form $a^3 = L \times B \times H$ and in the case of a sphere the measure will be $a^3 = 4/3\Pi r^3$. This was like comparing a triangle in relation to the half circle and the line.

It predates mathematics where the numerical use of determining a value was not yet established and only form was in use. It is equal to a time when we find in the half circle standing 180° related to the triangle (180°) and both still are equal to the 180° of the straight line notwithstanding the obvious differences used in form. However the starting point of these forms has to be equal and also has to be not zero to have the end be equal and result in all being equal in value in the end.

Kepler said a sphere is $a^3 = T^2 k$.

In honesty we have to realise that we cannot dismiss the whole formula that Kepler produced just because it doesn't match the scenario set to determine volumetric size as does the Newtonian version does. Kepler's version holds a foundation based on movement and it is in the movement we find the measure and not in the size as Newton's mathematical formula does.

In Kepler's formula the entire formula is formulating a circle being motion. However with the correct interpretation we find so much more than just motion. The formula is $a^3 = k / T^2$: That is what Kepler brought into civilization for all time to come. He saw space a^3 being in isolation due to the time it uses to move T^2 claiming such space forming independence according to the lines k indicate. Let us look at the factors in more detail before we proceed with the rest.

a^3 symbolises in a mathematical interpretation of implicating the three-dimensional space.

T^2 is representing the period or time that Kepler suggested we should use to calculate time that holds the orbiting planet in direct contact with the space in relation to a very specific centre moving from point T_1 to T_2 in relation to a precisely placed centre k^0.

k is the space taken from the centre to the end of the line k from which the planets must have grown if one accepts the Big Bang growth of particles and the affect of the Hubble constant on all cosmos material. The specific value about the centre is most important because from the specific centre gravity always apply the strongest influence.

The turning T^2 of any circle holding space a^3 is valid only if in reference k to a centre k^0.

Space a^3 will always be circling around as T^2 is in a position referring k to the centre k. That is what Kepler said when he said $a^3 = T^2 k$. Kepler indicated space a^3 will forever fight for independence and show separate individuality in remaining apart as identifiable cosmic components by means of motion. Every space will cling to independence indicated by k through fighting off the integrating of another coverall unifying unit by applying the motion of T^2! The problem we have to solve is what will the cosmos use to secure such independence between all particles? What sets space apart from the rest of space? First we have to admit that Kepler was the one that introduced the following.

Kepler gave us the answer to the following but no one ever took notice!

Kepler was the one that discovered **space / time** as **space** a^3 = **time** $T^2 k$

Kepler was the one that discovered **singularity** as $k^0 = a^3/T^2 k$

Kepler was the one that discovered **gravity** is holding **space-time** relative by the measure of distancing k as $k = a^3/T^2$ and $k^{-1} = T^2/a^3$

Kepler said gravity in space is about the area a^3 that would always keep equilibrium with the time T^2 it takes to travel the distance of the full circle position placed by the indicator k, therefore adjusting k as the need arrives. With k shifting in length a^3 will have to readjust and therefore T^2 will find a new relating value each time. This was the finding of Kepler and came after his intense study of orbiting planets.

Translating Kepler's mathematical expression $a^3 = T^2k$ correctly to the verbal statement in English Kepler said that there is a **space a^3** which is **equal =** to the motion in the **time duration T^2** thereof between two specific points which is a straight line **k** that holds a relation from a centre k^0 to an end **k** where the two ends run from the beginning of k^0 to connect at the end of **k.** I might not be the smartest boy on the block but I'm not that stupid either. I know how to translate mathematics into English… and I translate as follows:

a^3 must have a volumetric interpretation because the third dimension is sure evidence of multiple conjunctions of dimensions put together in three sides opposing three sides having the third dimension in place. The fact that any symbol uses a value to the **third power a^3** indicates **space** or a volumetric established and separate unit. Using a cube by three dimensions symbolises a cube, a room, a space to be filled, a unit able to hold other ingredients on the inside when empty or partly filled. It is space because it is volume using the third dimension.

T^2 is an indication of something having a cubic nature other than the square forming motion that is provided by the motion the square indicates, which is where the moving object is representing a third dimensional object that is moving from point to point and it is this point to point that multiplies into the square. The space is moving as a unit from one point to another point and the moving between the points are represented by a flat square or following a flat distance between two points. The cubic space was in one instant in one place and then the second instant in the other and because time can never stand still or become single dimensional (this I am about to prove) insisting that time must always support the motion it consist of or space as well as time in time cannot be. It is motion that is taking time, which is motion in the second dimension moving the space in the cube.

k^1 is the symbol used to indicate a straight line between two points with a definite beginning and a specific end position. It is the location where the form in question is holding space and where the space was and where the space are going to be in very next split instant that follows. This indicates points of representing **k** in different time positions to which the points will then be multiplying indicating form as a result of the square. The movement indicating not a square surface but movement by the square indicates the time the journey took to move the space from one point where **k** is indicating the location of the space to where the next indication of **k.** **T^2** will shift **k** where **k** indicates the space **a^3** that formed as a result of the movement **T^2** of being the space **a^3** indicated the point end of **k.** However, since time represents the square **T^2** and with **k** being the distance that prove that the **k** represents the distance the space **a^3** representing the form it is obvious that **T^2** represent the time that represents the space **a^3** in the square **T^2** through the motion. It is the distance moving space **a^3** in the cube to complete time in duration in the square of motion **T^2**; therefore **k** is permitted to be in the single dimension.

In the circle we may locate a straight line by reducing r that is symbolising the radius of any circle where such reducing will be indefinitely to the tune of halving r each time, then r would become infinitely small, even beyond human calculating means and become not a line, but a dot. However as mentioned in the case of the smallest dot holding one spot, r would become insignificant beyond human comprehension even, but never reaching zero and still Π would remain intact and dictating form. I believe one can begin too see where my suspicions are heading in my quest to locate singularity because the flaw comes about in the manner mathematics are practised for thousands of years and even today in our so called "modern times". The radius represents the initial line, the first that ever was. The very first instant in time was when Π appeared from infinity holding Π^0. With Π resulting from Π^0 the line will eventually become r^0 at a point ext to Π where the line began as a spot that grew into one dot and the dots added eventually to become a line. Finding this line made up of dots is most important when trying to decipher Kepler's formula.

Let us find the smallest possible line first. We already have reached the conclusion that by reducing the line, the reduced line will eventually leave all sides on the same spot on the condition that the circle spins. Such a spot must be round in form since it still holds Π as a factor next to r^0. We now are entering the domain of singularity where the visible is no longer traceable and only intellect can bring understanding of the scenario. With the line being the smallest line, such a line will start off as a dot Π that moved away from a spot Π^0. With all possible sides being in precisely the same spot we have all possible sides onto one spot. I chose to differentiate the dot and the spot by giving the spot a value of Π^0 while the dot holds Π next to r^0. Mathematically the spot is placing even form being Π in the single dimension Π^0 where the space is one (1) and holding exponentially zero (1^0). There the space moved over to form the spot Π^0 and by introducing form changed to the dot Πr^0 forming a circle as a dot.

Again I must draw the attention to the fact that we now are reaching into areas only the human mind can venture by understanding and seeing nothing more than with the eye of intelligence. The understanding of this concept demands our reaching the point where the mind of the animal cannot reach. If it starts with a line it then is there where that line only represents two sides being one and as such that is representing rather a flat Universe. At the dot Π we have roundness because we have Πr^0 while at the spot there is not yet any round form because of Π^0 and only when Π being round it then is requiring a shape or form and this lies beyond or before space at a time when any form of shape comes into the cosmos scenario. This part of the Universe came in a place at a time in a period where shape and form was a part of the distant future hidden in and beyond the developing eternity. The spot is located at a point that when entering the domain where the spot is located, it is also at the same time crossing the spot and landing on the other side where the radius becomes the diameter. Nothing can enter the allocated position the spot holds because entering the spot is crossing over to the other side of the spot. It serves us well to realise that the entire Universe was that small at a point where everything started forming because the spot that developed into the dot is still with every spinning circle...and the Universe is a multitude of spinning circles. It is also very wise to remember that once anything becomes a part of the Universe , it can never leave the Universe since it then has no place to go or no gate to pass through in order to leave the Universe . With the spot becoming a dot, there must have been a time when everything in the entire Universe was that big as the spot is, and that then moved on to form the dot and in that it went on growing in relevance.

There was a Universe forming in a time that everything present at this moment was so small that only form was in place and this was when the triangle, the half circle as well as the straight line was equal 180°, with no numerical values in place yet. At that point the line must have been so small it had reached a point not yet mathematically dividable in any way. The dot that formed was so small during that time that if at present any further dividing that took place, such dividing would have brought growth because there then would form space between the sides going in the opposite direction. However it is important to realise that anywhere we might locate Π^0 we also locate 1^0 because Π^0 is 1^0. The dividing brings all there is having all sides moving literally on the precise same spot, and I have located singularity in just such a spot.

I came to the conclusion that the spot I found had to be singularity purely on the grounds that that spot holds only one side to serve as a start to the starting point of all directions possible. In that side is only one spot where there is only one side applicable and one dimension present. In that spot space ended. That point is serving as a position for all possible points and cannot allow further dividing as it is in the smallest line or spot there may ever

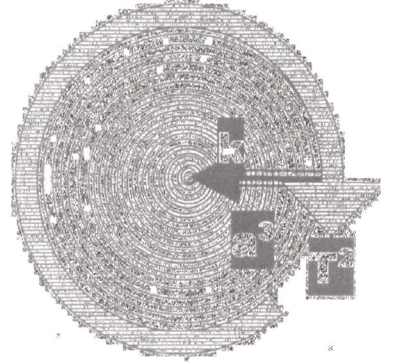

be. In the very centre of any and all circles spinning we find this point holding no space and therefore forming Π^0 that is 1, which is singularity.

Again I indicate the precise location of such a point. What is in the Universe, is spinning and therefore what I am referring to, applies to everything holding a place in the Universe and therefore this which I mention directly links everything holding any space whatsoever in the entire Universe. In the **precise middle** of all **objects in rotation** is a precise centre dividing the object in sectors that will **start the spinning initiation** from that centre point. Thus, the spinning object **will have a middle point**, a very specific **centre point that does not spin** and only holds Π as a specific value because no radius can apply. But also the one value such a line **cannot have is zero** because the line **is there and holds contact** to the rest of the material bringing about that **zero does not start any** line and therefore the **value of the line must be infinite**, just as described in **accordance** and by **the definition of singularity.** As I am introducing a very new idea, I wish to explain in better detail what I try to convey. While the toy top is spinning one will find singularity by moving the rotating line or radius progressively to the middle by reducing the length the line has from the edge to the middle. At one point all further reducing must end but the ending cannot include zero or nothing because the rest of the line still attach the rest of the top. As the rotating direction moves inwards, the rings will become smaller and smaller.

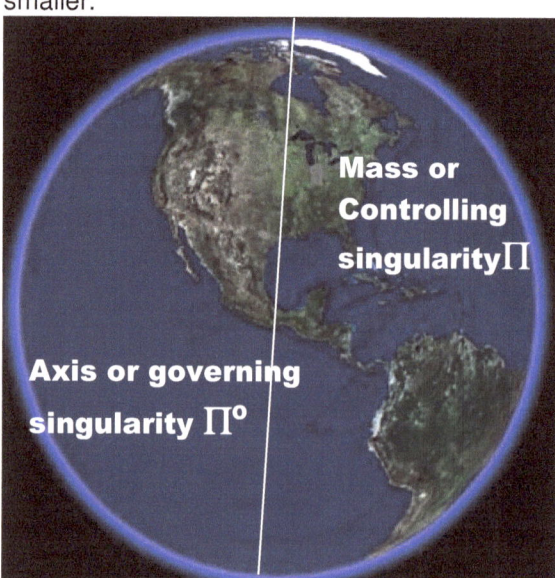

Mass or Controlling singularityΠ

Axis or governing singularity Π^0

When an object stands on the Earth the object is in mass because the Earth holds control over the atoms in the object. All the atoms spin around their governing singularity and all the atoms in the star connect to by bonding with the Earth's governing singularity. The Earth provides the centre around which the movement spins while the object with mass does the spinning as it then is part of the controlling singularity the Earth provides. The governing singularity of the Earth establishes the worth of the controlling singularity of the Earth by providing gravity or time in relation to the space moving. When an object falls towards the Earth the governing singularity of the Earth captures the atoms of the falling object while trying to establish control over the object falling. In that manner the atoms within the Earth forms the earth by becoming the controlling singularity in relation to the Earth's governing singularity.

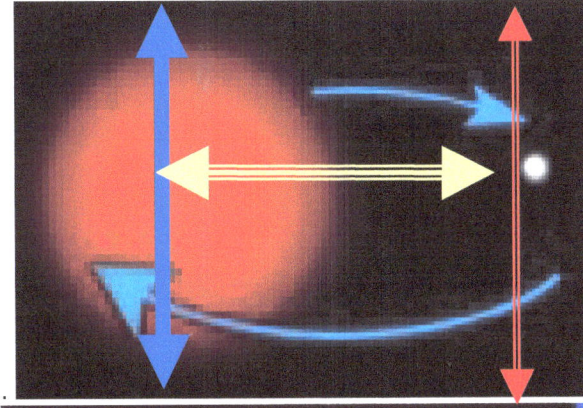

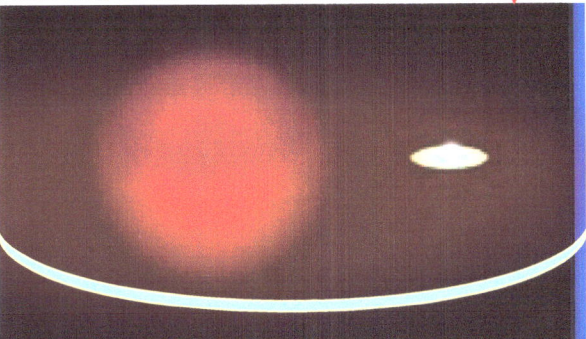

The Roche limit is the occupying of a dominant governing singularity that takes the control of the relevance of the controlling singularity that falls within the spherical dominance of a better developed star or structure. The governing singularity places the relevancy applying to the major star on the controlling singularity of the minor star which is the governing singularity of the atoms forming the minor star. As the relevancy can't form a constructive minor mutual singularity, the minor star structure conforms to the relevancy applying to the major star and with not enough material to form a bonding the minor star expands to the relevancy applying to the major star. We see the minor star expand to the size of the major star.

With the minor star spinning (sharing space-time) within the influencing range of the major star's governing singularity the governing singularity extends its control it has in the range of $\Pi^2 \div 4$ and takes charge of the governing singularity of the minor star. This capability puts the controlling singularity of the minor star also in control o the governing singularity of the major star making the entirety that formed the minor star part of the controlling singularity of the major star. With the governing singularity applying the terms of the controlling singularity, everything within the minor star becomes part of the major star since it is under the direct control of the governing singularity of the major star. The density applying in the minor star can't be confirmed or maintained with the spin that the atoms bring about in the minor star and the relevancy applying becomes subject to the terms and density conditions within the major star. Then by applying the Coanda effect the major star accumulates all that once formed the mutual singularity to become the mutual singularity that is part of the governing singularity control of the major star.

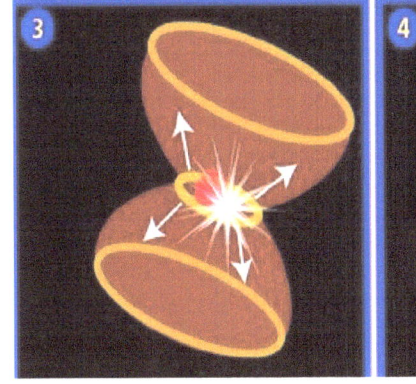

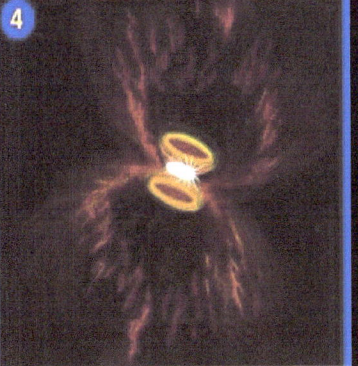

The governing singularity is controlling the atoms by means of placing the atoms in a relation where the atoms for the controlling singularity. In this expanding and the governing singularity losing control over the spin and therefore over the controlling singularity we have the Super Nova expanding the star's relevancy.

To understand what I just said we have to investigate singularity and where singularity hides. Singularity is Π producing Π^0 where this establishes Π and by movement Π^2 specifies space ending at Π^3. In order to locate Π we have to locate Π^0 and this we can achieve by producing movement Π^2 that would allocate Π^3.

All stars hold a centre point in singularity where that centre point has the value of being the equivalent of all the atoms where each atom holds a centre has an atom's worth that combines to form an equal to the value equal that the star's singularity is worth.

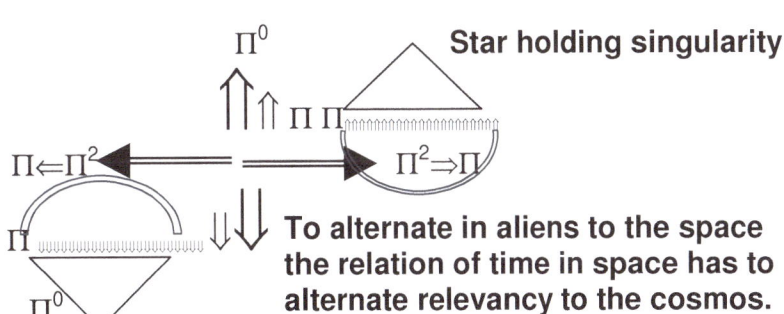

Star holding singularity

To alternate in aliens to the space the relation of time in space has to alternate relevancy to the cosmos.

The point is in everything that rotates and everything in the Universe combines to form a connection that connects everything to all other things.

That point albeit hypothetical, is also as much a reality none the less and is placed where that point **must be standing still** because every line **running from that point** in **opposing directions** is also **in opposing directional spin the other or opposing side.** In considering the spinning motion in the fraction of time in the detailed instant every aspect of rotation will turn in every instant of change in time. Although the points had the same characteristics only one instant before, they oppose the characteristics it had just before and just after the very instant in which they are and to which they relate by similar points also in rotation. The fact of the graph proves my point in quarterly opposing dimensions and values.

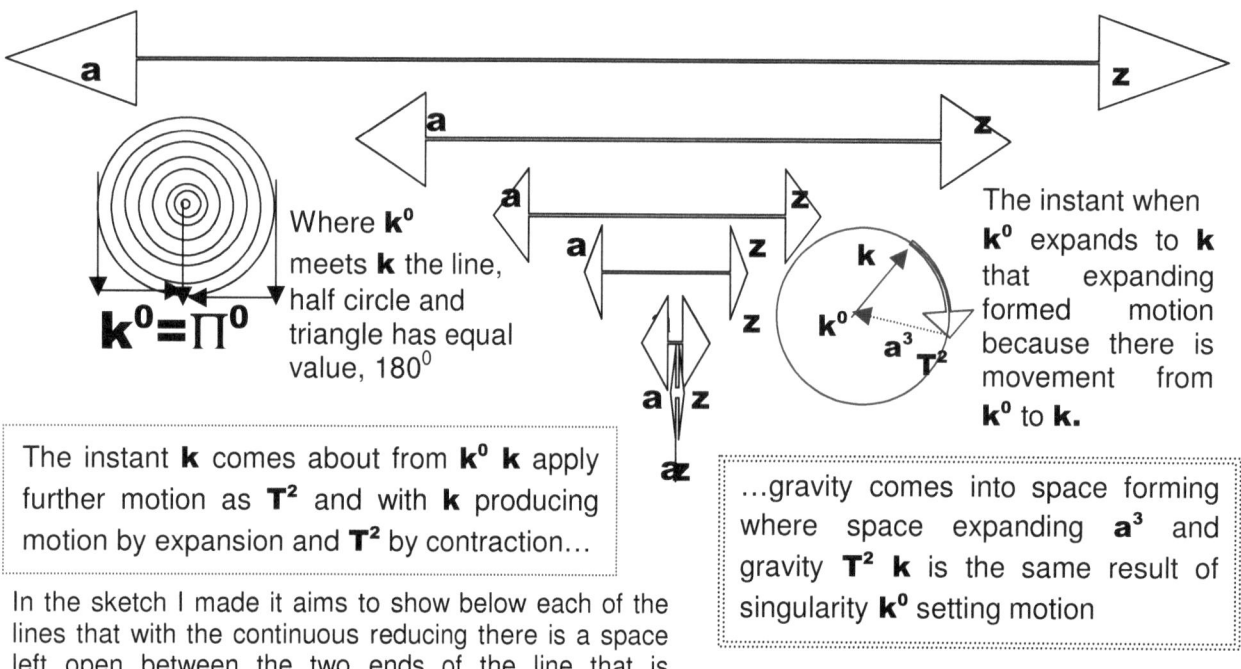

Where k^0 meets **k** the line, half circle and triangle has equal value, 180^0

$$k^0 = \Pi^0$$

The instant when k^0 expands to **k** that expanding formed motion because there is movement from k^0 to **k.**

The instant **k** comes about from k^0 **k** apply further motion as T^2 and with **k** producing motion by expansion and T^2 by contraction…

…gravity comes into space forming where space expanding a^3 and gravity T^2 **k** is the same result of singularity k^0 setting motion

In the sketch I made it aims to show below each of the lines that with the continuous reducing there is a space left open between the two ends of the line that is symbolising the end of the line in reducing. In the very end where numerical value becomes infinite the two ends will share one location even by having one single point holding each one. There is no chance that I can present any sketch reducing the line to a point where the points are sharing one location literally in the single dimension. Yet although sharing one spot the points are there and with the points being present they may not be dismissed as nothing. From there no reducing in a natural manner can lead to nothing without changing the rules of mathematics in such reducing. But the two ends has reached a position where any further effort of reducing must bring about the start of extending the line once more because every point possible share space with every other possible point at the point of singularity where all points share one common space.

By moving any of the points, such moving by further decreasing at that point must then bring about an increase of space once more since the space at that point in the centre spot of the circle spinning has gone infinite. This also applies to the sphere that is a multitude of circles because the circle uses a line to indicate size running from a centre to an edge. By reducing the line and by reducing the circle the reducing will end up having the ends in the same position in the very centre of the circle. It is this fact of the moving in any direction of any point from that spot holding singularity that such motion will introduce space as the space exceeds the previous limits of singularity. What I am trying to say is by moving from the spot Π^0 to the dot Πr^0, such movement evokes Π^0 a spot to Πr^0 forming the unseen line and without the movement of the spot Π^0 to the dot Πr^0 to form the line Πr^0 the allocating or positioning of singularity Π^0 will not take place.

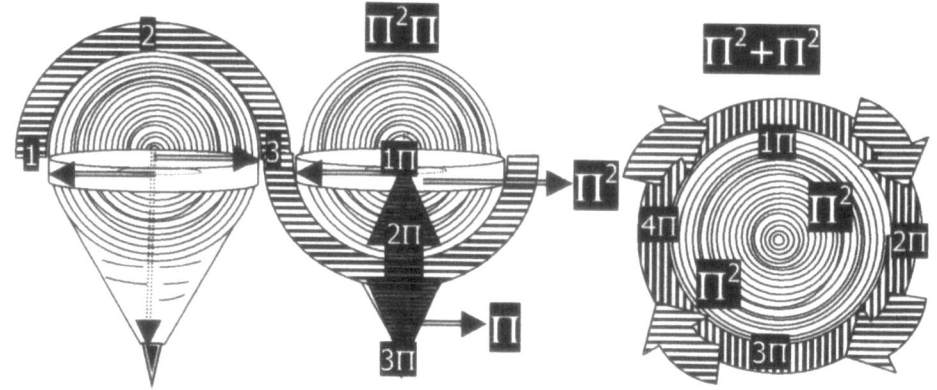

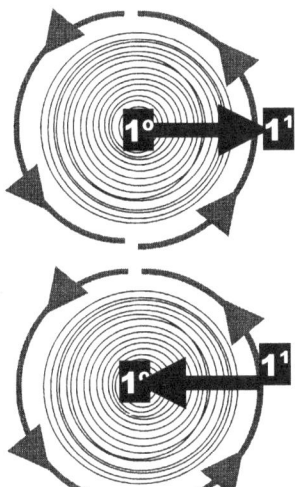

In the Universe there are one substance forming the entirety and that is singularity. Since singularity holds no space it is not a reality but is only in place functioning in terms of all other points holding singularity that spins. Therefore the crucial connection in terms of singularity is found in the movement applying. There is one type of singularity, which is one, but there are two forms of singularity where the one type spins in control of and the other spins with no seeming control applying except time serving as the control by producing the applying movement also known as the Hubble constant. In that I named the one being condoled by spin 1º and the other being controlled by linear expanding through time 1¹. The one is material performing in relation to what never is able to spin but holds control over everything by that which can never stop spinning and in that that which can never stop spinning serves as the controlling singularity to that which can never spin that becomes the governing singularity.

The spinning established a value between the object's movements in relation to the movement of the electrons spinning around the atoms. The evidence that proves the statement is the fact that the minor star relinquishes density when the major star takes control of the minor star's governing singularity and thus takes control of the movement making the control of the atoms dependent on the major star's governing singularity.

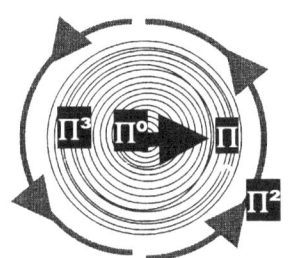

The movement of the atom is interlinked with the movement the star establishes which we call gravity. This movement depends on the spin but also depends on the movement of the structure in its rotation of the secondary controlling object as the Sun is in the case of the Earth. This is in relation to time applying to the atoms versus the space that is either dismissed or displaced. The movement is as all movement is, space – time incorporated.

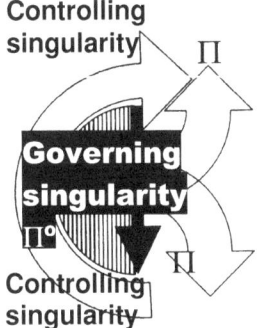

Controlling singularity

Governing singularity

Controlling singularity

In the spinning top we find that singularity Π^0 can be generated by motion. But singularity Π^0 has no motion within the dimensions we find allocated to the Universe in which we live. Since the singularity found in the centre of the spinning top is in truth just a mathematical point, which means in mathematical terms the point with no sides cannot even be calculated as a factor since the measure thereof goes beyond what mathematics ever can calculate. Mathematics has a use within the 3D Universe but singularity that keeps the spinning top attached to singularity governing the gravity of the Earth, that singularity is truly single dimensional and beyond mathematical measure. It is singularity Π^0

If we put this in terms of singularity (Π^0) we find the Earth (Π^3) is in relation as viewed from Alfa Centauri (Π) four point six years (Π^2) while moving in that space that is time that has gone by. That secures the three dimensional status the Earth has (Π^3) in terms of a present (Π^0) that depends on a location (Π) secured by a future (Π^2) that will come by movement where the future ($\Pi = \Pi^3 \div \Pi^2$) moving forward that also doubles as a past ($\Pi^{-1} = \Pi^2 \div \Pi^3$) by the light coming from and thereby confirming the past. That is space formed three dimensionally by keeping time in infinity apart from time in eternity. The relevance (Π) that forms in relation to the present (Π^0) will relate to movement (Π^2) and the movement is circular which ensures that the relevancy forming is circular (Π) by securing that the movement is circular (Π^2) in terms of one specific point (Π^0) in infinity which then secures a roundness (Π^3) that forms an everlasting eternity ($\Pi\Pi^2$) which validates a never ending circleΠ^3. In this time in infinity (Π^0) that secures that there is an everlasting eternity ($\Pi\Pi^2$) in space (Π^3), it is not the space that is everlasting but the movement of time by the line ($\Pi\Pi^2$) that is everlasting. The **governing singularity** (Π^0) holds a **positional validity** (Π^3) of three dimensions $\Pi^3 = (\Pi\Pi^2)$ in terms of any **relevance** (Π) formed by the **controlling singularity** ($\Pi\Pi^2$) thus mathematically it equates to $\Pi^0 = \Pi^3 \div (\Pi\Pi^2)$. If a **relevance** ($\Pi$) did not validate a **positional validity** (Π^3) securing a **governing singularity** (Π^0) in terms of movement formed by **the gravity** (Π^2) that produces the **controlling singularity** ($\Pi\Pi^2$) in space, with a three dimensional status Π^3, then space (Π^3) would not be obtained and thereby the Universe would not be secured. That is why space-time is $\Pi^0 = \Pi^3 \div (\Pi\Pi^2)$. However this must be seen where it applies. It applies where singularity as time meets space, which means it applies at a point in the Universe where time still grows and that is at the position that predates the Big Bang. It is where material forms before material forms. It is where the visual will never come. It is where singularity Π^0 forms space Π^3 by singularity (Π) moving (Π^2). This means there is a time space delay validating a connection where any connection is (to us) absent. Stars in an axis dispute are not yanking each other around because mass is pulling one another because of some magical medieval territorial dispute. There is control over atomic control coming as a result of singularity charged by and charging movement. by virtue of a controlling singularity effectively in place due to a governing singularity much the same way as a top spinning tries to surge into the air when spinning too fast or try to fight of the control the Earth takes on the top just before the top is grounded by movement deprivation coming from the Earth.

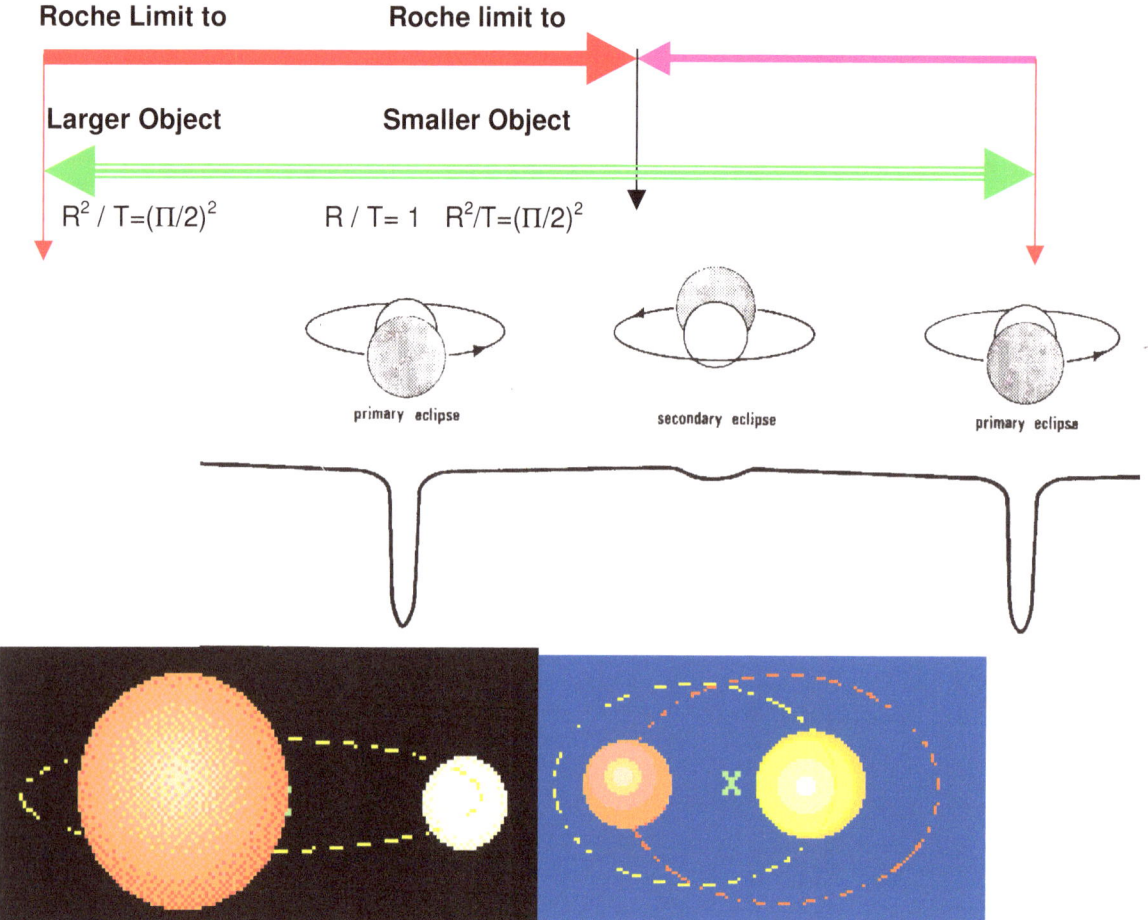

This is the reason why stars would form massive binaries, where they share a common combined circular displacement or a controlling singularity in terms of a governing singularity, separated only by each stars Roche limit with no linear value. As the Titius – Bode Principle comes into effect the linear displacement would once again grow, or the common spin value will be to grate for either one, or both, and their structural composition will collapse, forming smaller structures with less space to occupy the time in which they are. This proves that atomic movement shows loyalty to the control of a governing singularity that forms a president in movement and a governing singularity takes charge of the entire atomic movement.

The governing singularity of the one star takes charge of the other stars movement and thereby the controlling singularity which is also taking charge of the mutual singularity and in that it takes charge of the individual atomic singularity as the star deprives the other star's atoms of any independence adhering to the atomic governing singularity in the star. It is this evidence by which we can gauge that singularity cross refers and establishes space-time worth in not only the star forming the mutual singularity but also the star of which it took charge of the controlling singularity.

Where the one governing singularity can't takes control of the other governing singularity a fight proceeds where movement is detrimental in which star takes control of the next star. This has nothing to do with Newtonian mass. The one star has atoms spinning around a governing singularity and all of the atoms forms a spin that vests a value into the star's centre singularity. Because the atoms are 1 and the centre is 1 the atoms is not only equal but the atoms' centre is the star's centre and

Important to note is in the case of Binary stars the two stars "lock out" space-time. By finding where it all began is equal to finding where the line began we have to trace the line in order to trace the development of the entire Universe. As seen the first development went beyond where mathematics may take us. The Universe did not become more but only focussed better on detail. What is was present because nothing can be new to the Universe that started out to be what it presently is.

Governing singularity

Individual singularity

Individual singularity

Mutual singularity

Mutual singularity

Controlling singularity

Controlling singularity

Principle singularity

An object can rotate in outer space as maintain a speed that will keep the straight because part of all lines in the relation between the factor of how much **k** influences or how much T^2 influences and the combined unit determines a^3.

long as it can object rotating in that orbit. No line can go cosmos is the curve. There is always some

That reason too has its footing in the Titius Bode law because 7/10 is half of 10/7 and by going seven the factor also has to go 10. The compliment of this is gravity at Π^2

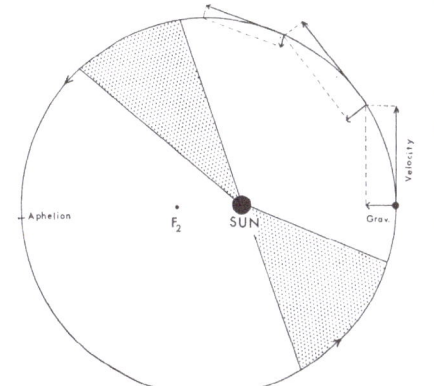

One of the four most important values in the Universe is the Roche limit. On this and the Titius –Bode connecting to the Coanda effect by means of the Lagrangian points and on the inter connecting of the four principles rests the growth of the universe, as space relate to time. The sound barrier is the four cosmic principles applying to form the relevancy between the governing singularity and the controlling singularity.

The Roche limit comes into effect when the linear displacement factor reaches a value of one and part of the circular displacement value. In this is the value $\Pi^3/\Pi^2 = \Pi$. When two stars are at the Roche limit, the linear displacement reaches a value of the lesser one, and excites in an electrical sense the singularity extend the atmospheric level to

within the lesser one charging its atmosphere to that of the major one.

$$\Pi^3/\Pi^2 = \Pi$$

$\Pi^3/\Pi^2 = \Pi$
The circular displacement reaches The roche limit forms a centre by and the atoms all foming a structure holds a unity by measure atoms by.

$(\Pi/2)^2$

its full complement of half Π^2 which is the Roche limit. the value of the atroms spinning inside the structure governing singularity secures independence while the of the governign singularity that the star controles the

Atoms spinning inside the structure independently form the individual singularity

Atoms inside forming the total structure is the mutual singularity

Inside centre forms the Governing
•
singularity

Curvature formation spinning as a structure entity forms the Controlling singularity

Spinning around the Sun produces the principle singularity

This indicates four factors forming singularity that absolutely dictates the cosmos in terms of movement. Holding that in mind, I therefore had to name the four positions that equally form singularity by dictating gravity. To argue this concept of singularity guiding movement, let's take the Sun that provides a centre k^0 for the Earth a^3 forming a centre where **k** points a line that forms the orbital circle T^2 wherefrom the edge of the line **k** is pointing at the position of whichever planet a^3 forms a circle T^2 in relation to a line coming from a centre of the Sun k^0. The line **k** indicates the distance from the Sun's centre to the planet that orbits and this forms the circle as the planet a^3

orbits T^2 around the Sun. The line **k** will provide a line from the Sun's centre k^0 and the line **k** will provide a spot where T^2 produces a circle holding space a^3 in a located position by running around the centre of the Sun k^0. In this view the space a^3 of the Earth rotates and in that forms the <u>**controlling singularity**</u> that holds the value as Π indicated by **k** forming between **k** and k^0 being singularity Π^0. The Sun holds singularity in the centre, which is forming the <u>**governing singularity**</u> Π^0 and from that point the circle T^2 comes that forms the orbit Π^2. That means every single point that **k** indicates there are positions forming space a^3 implicating sides of a double dimension. In the same manner is **k** not limited to distance or is T^2 lesser by size. If Kepler said $a^3 = T^2k$ then $k = a^3 / T^2$ is also what Kepler said. There are three dimensions a^3 between any two points T^2 flowing as time from the centre of the Sun, which is indicated by the line **k**. However in the next scenario the Earth holds the <u>**governing singularity**</u> Π^0 running from the centre k^0 to **k** forming the edge while the circling rotation T^2 then forms the <u>**controlling singularity**</u> Π indicating the point in rotation. There are also two other points holding <u>**the mutual singularity**</u> and <u>**the primary singularity**</u>, both which I do not explain in this presentation but without which the four phenomena would not form gravity.

The value of **k** is not to be put in place as a measured value, but is there to bring a reference to the location of singularity $k^0=a^3/(T^2k)$ applying as to place a specific singularity in as the <u>**governing singularity**</u> and acknowledge the position of another singularity in place as the <u>**controlling singularity**</u> because there always has to be a <u>**controlling singularity**</u> determining the orbit while there has to be a <u>**governing singularity**</u> determining the spin of the body in relevance performing as the space a^3 in question in the formula $a^3=T^2k$ where in that formula **k** determines the relevance of k^0 as in $k^0=a^3/(T^2k)$. However, this burdens **k** forever with the responsibility of forming a line and a line is what places the Universe in place while the circle T^2 is forming the Universe a^3 at the same time. Every space a^3 in question puts singularity k^0 in position by the motion T^2 in relation **k** to the position allocated to **k** in the Universe a^3. Nothing in the Universe can move without moving straight **k** that is also going in a circle T^2 to form space a^3 in relation to a centre k^0 while in orbit around another centre k^0. In this point k^0 time forms space and space develops as the history of time running from k^0.

a^3 symbolises in a mathematical interpretation of implicating the three-dimensional space holding a specific centre in relation to another specific centre indicated by **k** that could apply to either centre points in question. This is always a straight-line **k** representing the position of the <u>**controlling singularity**</u> moving in a circle T^2. The space forming a^3 is a **positional validity** of the space indicated by $k^0 = a^3 / (T^2k)$.

T^2 is representing the circle that goes around the <u>**governing singularity**</u> k^0 or Π^0 that forms in relation to the line **k** pointing to the controlling singularity or Π in reference to the centre k^0. The space that forms holds the orbiting planet a^3 in direct circular contact with the space in relation to a very specific centre k^0 moving from point T_1 to T_2 that then forms Π^2 in relation to a precisely placed centre k^0. The circle coming about from T^2 is the <u>**controlling singularity**</u> Π, which is always a circle Π relating to the centre Π^0 that is positioned by the line **k** in relation to the centre k^0 and by forming a circle Π it holds reference to the <u>**governing singularity**</u> Π^0. Where <u>**the governing singularity**</u> is the centre of a spinning object such as the Earth, the centre of every atom holds <u>**mutual singularity**</u> Π^3 that collectively puts a mutual value of all the atoms' singularity as a combined equal to the <u>**governing singularity**</u> Π^0. The solar system will provide a **primary singularity** $\Pi^3 = \Pi\Pi^2$. The one would represent T^2 the other forms **k** that then produces the third singularity forming space a^3.

k indicates <u>**controlling singularity**</u> from the centre k^0 ending at the line **k**. This line shows the location around which a planet circles. The specific value about the centre is most important because from the specific centre gravity indicates a positional worth. The line forming **k** is pointing the circle or the <u>**governing singularity**</u> formed from a line that ends at a circle T^2 running from the centre k^0 to where the space a^3 is indicated.

The turning T^2 of any circle holding space a^3 is valid only if forming a reference **k** to a centre k^0. $k^0 = a^3 / (T^2k)$. This depicts a position the domineering singularity k^0 fills in relation to another point serving subordinate singularity **k**. There are always a dominant and a serving singularity interacting. If **k** indicates the centre of the Earth then T^2 rotates initiating the <u>**governing singularity**</u> k^0 where then the centre of the Sun **k** will form the <u>**controlling singularity.**</u> When the Sun rotates, the Sun's centre k^0 forms the <u>**governing singularity**</u> giving the Earth in orbit **k** holds the <u>**controlling singularity**</u>.

The measure of **k** is not a specific value but serves only as an indicator to which space rotates or applies by the space rotating in a circle. This role of singularity being <u>**controlling**</u> or <u>**governing**</u> is playing part in movement of gravity forming and is very important when trying to understand the role that the four phenomena play in forming gravity. It is important to understand what happens in the event of an object going through the "sound barrier" or when escaping from the Earth's atmosphere. Where the object is standing still holding a position that allows the object to have mass, the object is part of the Earth while the Earth has the <u>**governing singularity**</u> and the Sun has the <u>**controlling singularity**</u>. As soon as any object moves on Earth, the movement switches singularity by allowing the object to obtain the <u>**governing singularity**</u> while the Earth then for fills the directional circular control in forming the <u>**controlling singularity.**</u> All four phenomena interacts in a manner forming this role where for instance in the solar system the Sun holds the <u>**controlling singularity**</u> and Milky Way forms the <u>**governing singularity.**</u> To find validity in my argument one must draw this statement of motion back to the point where singularity is getting sides or said mathematically Π^0 is going Π. Π is the **controlling singularity** and Π forming Π^2 is in relation to the **governing singularity**Π^0. When there is singularity there can be no sides. The one forming singularity Π^0 by measure fills no space while form Π develops Π^2 into space. The space that even the dot fills being Πr^0 does not really exist in the manner we humans see space to exist. It is a spot that is there without being there. It does not visually exist because it is not filling any substance and it cannot be recognised since it is not three-dimensional. The spot and the dot have no dimensional worth of any measure but holds relevance. This Universe I am addressing has never been unveiled by any one since this is the flat Universe. This Universe holds a line in time made up of dots and spots forming no space but holds a Universe relevant.

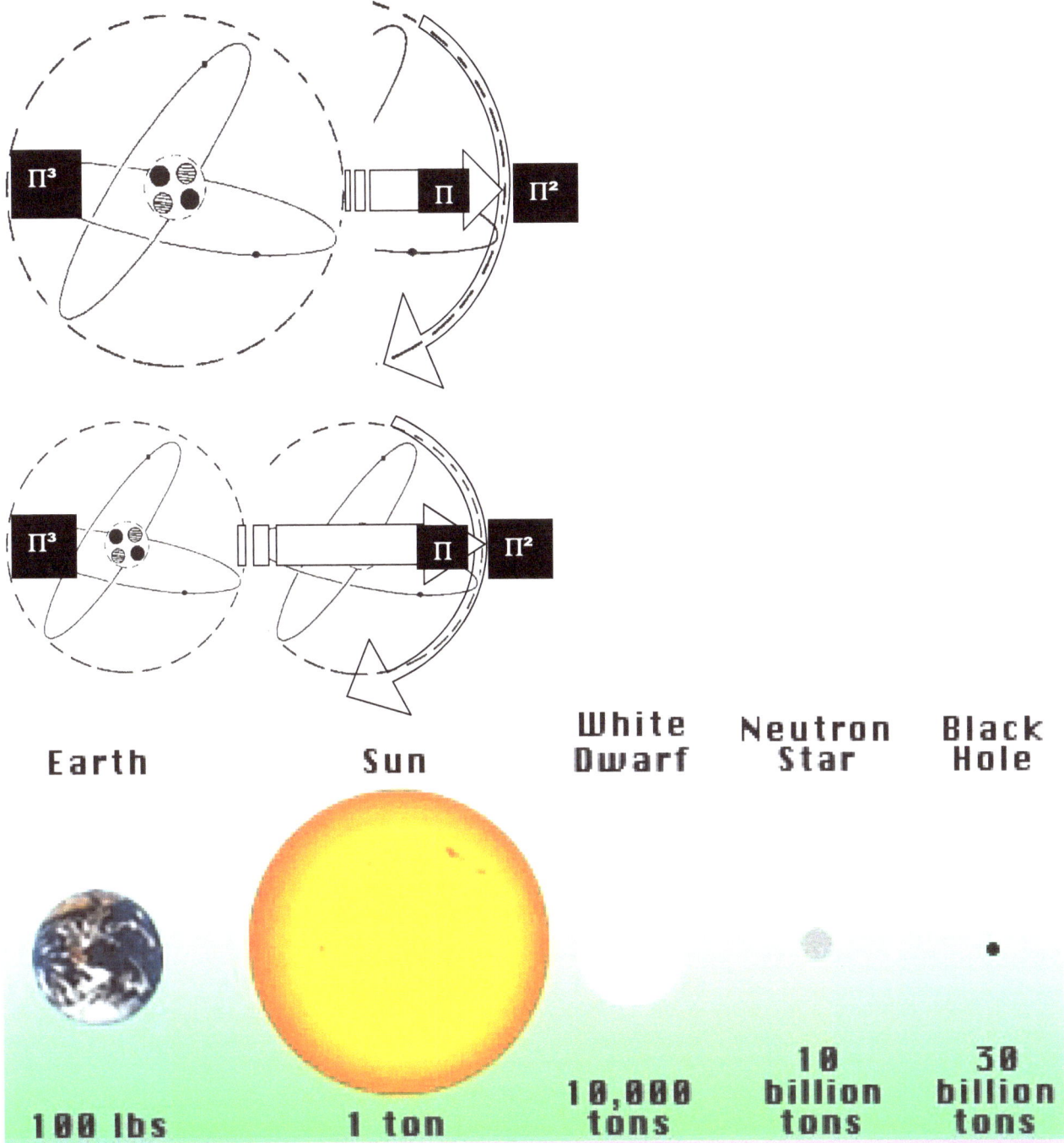

In every case scenario as gravity intensifies the atomic space becomes smaller as the density factor rises in every case and the mutual singularity becomes more compact because of a firmer controlling singularity being the product of a more intense governing singularity.

In determining this behaviour as part of a cosmic process where matter interact with matter in an laid down set of rules, once more we should be asking questions and this time it is whether the top will show the same behaviour in outer space as it does on Earth. With the reply of no it would not come as an admitting that the process involves the interacting of singularity of the Earth with the singularity of the top where the spinning created independent singularity, is as valid as that of the Earth because the Earth has a role in sustaining it or destroying it at the border ends.

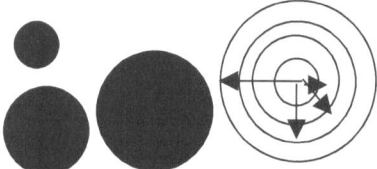

Looking at the affect of gravity it shows the precise quality of no distinctive point, as gravity never seems to end at a point but flows all over affecting all that holds a position in its sphere of influence. The gravity coming from China meets the gravity coming from America at no particular spot but intermingles without distinction.

Using the concept that gravity applies Π as the circle factor Π as well as Π^2 replacing r^2 the replacing by Π brings two values as Π and Π^2. That I found is the case with gravity and will be apparent when explaining the sound barrier as well as the Four Cosmic Pillars. In order to create a distinction I remained using r as the indicator of the cube or non-circle that has vacant space and by vacant space I refer to non-solid structures. In the solid structure I use Π as a value for reasons that will become apparent in due time.

This spot is the result of a most basic process of reduction as the Hubble constant is a most basic process of expanding during a matter of time. By reducing the line constantly the only value that will eventually remain without dispute from any party arguing about the facts is one followed by an exponential zero (1^0). By only having exponential zero instead of a numerical zero and a radius as one in the square (the radius effectively becomes one holding any and all sides on one point) such a point might become any value of any significant measure implicating anything but zero as the radius. By expanding the line, it will be an evenly spaced structure growing into the most perfect round dot ever possible anywhere at the point when it starts to grow.

The reducing of the line is one dimension in six and although such reducing is representative of two indicators all the other indicators must still be accounted for two. In mathematics there is a line being one quantity and the circle indicator Π being the next circle indicator. Reducing the line will erode the value of Π by ratio. That will eventually lead to having a circle ratio of Πr^2 and eventually lead to Πr^0 but that is not the point where the circle ends. That is where the ratio applying factor ends but it cannot exclude the circle. The circle as a concept can still reduce when it abolishes form to the single dimension. It is not the radius that is responsible for the circle but the figure value of pi and by abandoning π only then does all the aspects fall back into the single dimension.

The circle can reduce one step more when the circle eliminated r completely by returning r to a point of singularity r^0, but the elimination of r as the factor reduced the major factor to the single dimension in Π^0. That will not reduce the cosmos to zero, but it will only eliminate all potential lines r^0 to potential circles $\Pi^0 r^0$ and from there the circle Πr^0 will come about by manifesting as a line but that manifesting can firstly only establish a circle Πr^2.

The only value that singularity can have although the single dimension may host the entire Universe is Π^0. Pick a number and elevate it to the power of zero and in the process one may have established another point holding all points in singularity because that is the value of singularity. Only Π^0 or any other value holding one accompanied by zero as an exponential value can ever be the accurate value to singularity while singularity will then host the rest of all the possibilities in the Universe.

This means that the entire Universe composes of and is made up of singularity... this much I am going to prove. Every point occupied or otherwise constitutes of singularity either under control by movement in a form we call atoms or being passive in a location we call outer space. I wish to repeat the position holding singularity because if I introduce anything new, then this centre singularity is the pivot of everything that I introduce to science, and also I refer to the top because Newton used the top as an example by which Isaac Newton missed everything that Kepler clarifies about cosmology.

In the sketch to the right above the circle to the right would come about from a straight line r growing influencing the appreciation of Π, but to influence Π would lead to a breakdown in r as Π and r are different entities. The circles to the left (black dots growing in size) shows a continuous growth by extending Π every time and since Π is the same part as the previous Π, only extending that billionth of a millimetre each time, the circle will be truly continuous without any signs of a break.

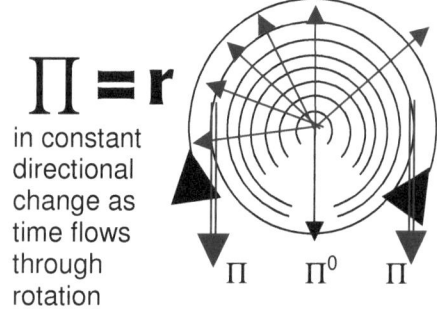

$\Pi = r$

in constant directional change as time flows through rotation

Π Π^0 Π

The new direction pointing to a new location in relation to the previous point will oppose the previous point it had in relation to direction considering the centre point.

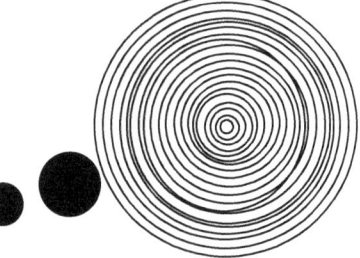

Let's go back once more and reduce the line by half every time. Then repeat the process until it can repeat no more. The reducing of the line by half every time will get to a point where all the ends land on the same position without any possibility if halving the two ends further. The points share one position and moving the points in any direction will lead too an increase of the line once more.

Locating and finding Singularity

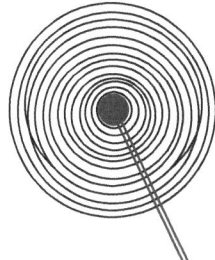

The entire Universe consists of lines running all over.

In the **precise middle $k^0 = a^3 \div T^2 k$** of all **objects in rotation** is a precise centre dividing the object in sectors that will **start the spinning initiation** from that centre point. The object has to rotate $T^2 = a^3 \div k$ in order to instate the space $a^3 = T^2 k$. Thus, the spinning object **will have a middle point k^0**, a very specific **centre point $k^0 = a^3 \div T^2 k$ that does not spin** and only holds Π as a specific value. One value such a line **cannot have is zero** because **zero does not start any** line and therefore the **value of the line must be infinite**, just as described in **accordance** and by **the definition of singularity**

That point albeit hypothetical, is also as much a reality none the less and is placed where that point **must be standing still** because every line **running from that point** in **opposing directions** are also **in opposing directional spin the other or opposing side.**

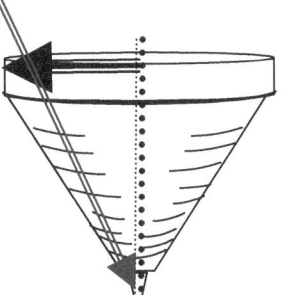

In considering the spinning motion in the fraction of time in the detailed instant every aspect of rotation will turn in every instant of change in time. Although the points had the same characteristics only seconds before, they oppose the characteristics it had just before and just after the very second in which they are and to which they relate by similar points also in rotation. The fact of the graph proves my point in quarterly opposing dimensions and values,

From this centre line that is only theoretical definable, but is still there all the same, an centralised line forms holding opposing values apart and parting the opposing values is what proves that this line forming has no status in space and yet controls all that holds space. The one side will turn left and by crossing this line holding no space, all the space will then turn right.

The parting of directional opposing space will always form what becomes real and distinct when rotating, and by rotating around this line that is only theoretical the spinning is putting this line not rotating even in more distinct prominence. Because of the rotating that evokes the presence of the line not being there, the influence this line holds grows so much it covers all the matter from end to end, to a securing the spinning to a point in the centre that holds no spin value.

It is there as well as not being there by not spinning because it has no space to spin. Yet the line is there and therefore it has the most original value anything can have. When not rotating, the line disappears and only a diameter runs across the material with the diameter going as thick as the material will go.

When rotation begins, the line then forms according to a radius. While the line forms where the radius starts, it shrinks back to a hypothetical position claiming zero spin that through that it is not less distinct but more distinct because from that point every rotating becomes a piece of what ever forms part of what is then spinning. Because it spins the end of the line forming will clearly carry the singularity value of Π^0 to end at Π that then is implicating rotation by the value of Π^2.

When looking at the cosmos from whichever angle it indicates the fact that the cosmos is moving. Everything in the cosmos is moving and all things about the cosmos are moving in relation to everything else in the cosmos. Everything is forever spinning in relation to a point that could be any point that is not spinning and everything is all going towards as much where it is coming from.

Everything is on the move and always encircling something of making that centre point to seem to be of greater importance than what everything is that is spinning. A top can spin but the parameters of its spin are limiting the motion it can apply. By not spinning the top is still spinning as the Earth is doing the spinning on its behalf.

The spinning top that Newton dismissed as $\frac{dJ}{dt} = 0$ brings all the evidence any one needs in order to come to a conclusion that will bring any proof that the singularity governing the top connects too everything anyway. Placing singularity is fair and fine, but what will the evidence be in proving its activeness as part of the creation at large? The reason why we can be sure it is active is that when spinning it shows borders implicating restraining of further movements outside the set limits. By going faster (past the upward border) the spin goes oblong where it actively tries to change the position the top holds to the Earth in relation to the surface of the Earth. By going too slow it once again shows identical characteristics. When going too fast it indicates an attempt to rise into the air, therefore relieve its singularity in an effort to part with the Earth's singularity. It shows unmistakable characteristics of trying to become airborne securing an independent position from the Earth, which holds it down.

Pinpoint positioning of singularity Π^0 with Π positioning space to either side forming the border set by singularity

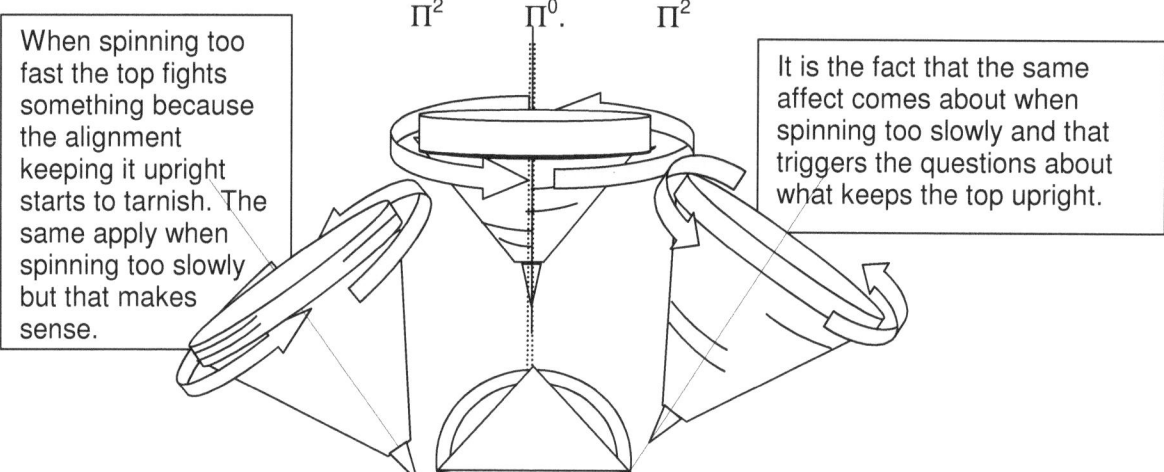

Π^2 Π^0. Π^2

When spinning too fast the top fights something because the alignment keeping it upright starts to tarnish. The same apply when spinning too slowly but that makes sense.

It is the fact that the same affect comes about when spinning too slowly and that triggers the questions about what keeps the top upright.

The rising above the position the Earth holds the top is clearly indicating the top is trying to generate more gravity as what the mass would be by which the Earth restricts the top and hold it in a position where it will form mass and then become part of the material forming the Earth. When it is at the bottom we surmise correctly that it wishes to topple over and fall down, but something drives the top to put up a fight in order to stay upright. If the top falls down, this action of falling down will kill the centre line holding singularity and it is that centre line the top holds that is keeping the top erect. By destroying the line it will then enforce stopping the independent spinning of the top. Of course the bottoming out shows the same characteristics whereby we gauge that to be the normal process of falling down. If the bottoming is relative to the Earth's singularity and we recognise the process as normal, then the top of the limits should be just as recognisable normal.

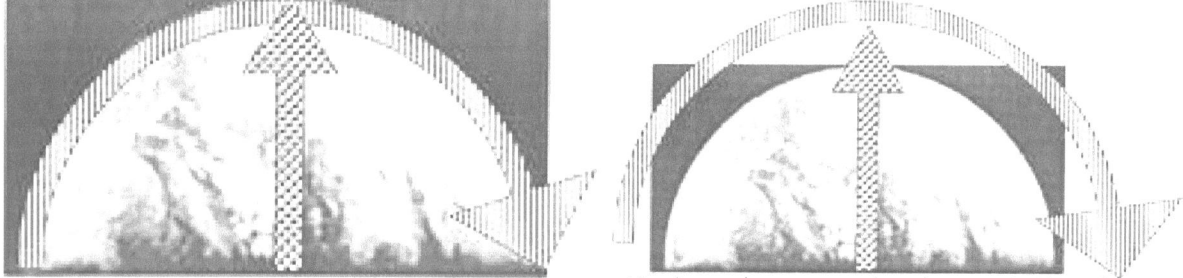

When any object is in a state of having mass the object has to be standing still and being secured in a position on the Earth at that point of having mass. The object has to be in a position of absolute rest while it is on the Earth. At a point of standing still in relation to the Earth while excepting only the movement the Earth allows any object to form mass and it is where at that point that the object with mass is resting while all the rotational movement is equal to the movement the Earth delivers where the Earth is rotating.

Rotating at the speed the Earth dictates form the factor science call mass. When the object leaves the surface of the Earth such an object will have to move much faster than the Earth moves or have less density than is required to maintain a steady position on the Earth. When any object is standing still being in a state of having mass on the surface of the Earth, an object has micro gravity because the individual gravity left to the object in mass is infinitive small and is left to become an indication of attempting further movement towards the centre of the Earth while the Earth's material blocks the micro gravity to move and hence apply mass in doing so.

Mass is not something inherent of the object but is the annexing of the object given mass by the Earth to secure the position of the object to ensure the object becomes part of the Earth structure. Having micro mass (not micro gravity) is where the body in rotational movement extends beyond the limit at the point where the Earth surface would award a mass factor. The movement speed goes beyond the speed required by the Earth at the Earth limit where rotation velocity secures mass as a factor. By exceeding the rotational velocity at a higher rate, such movement would exceed the movement or gravity of the Earth that is required in order to grant a mass value.

When the top is spinning it is this line that urges the top to excel from the limitation of the gravity of the Earth and extend up into the air and away from the ground. It is this centre line holding singularity that drives the top to lift up from the ground and fight the mass that the Earth inflicts as to retain the top with the limitation of enforcing the top into a state of mass where the mass holds the top onto the ground. The top is fighting the Earth's effort in restraining the top with mass by producing gravity that lifts the top into the air. It is the top's spinning that is producing anti gravity to fight off the gravity of the Earth. I have heard so many scientists refer to man discovering anti gravity as if such a discovery of a force of anti gravity will give humans the power only God can have. It is this mindset that I refer to as science wanting the marvellous, the magical and the unexplained. To the masterminds of science having anti gravity would come to the same as unlocking all the witches' forces and opening the Pandora's Box of forces while anti gravity is simply jumping in the air. If gravity is what is pulling you down, then anti gravity must be something lifting you up.

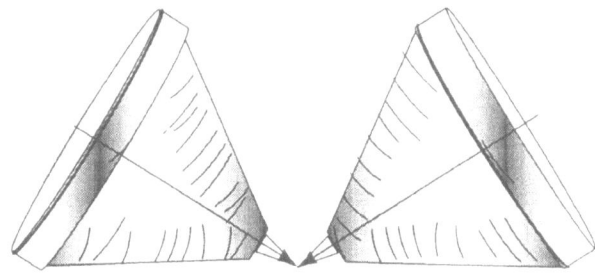

When the spinning has died down so much it will arrive at a point where the gravity of the Earth will reduce the spin of the top to lying still while the Earth secures the top with going into having a state of mass or having no independent movement which is having mass. It is at that point that the centre line within the top that is securing gravity by the spinning of the top will seize to be and the top will once more come to rest in a state of mass. The gravity of the Earth is fighting the gravity of the top which is equal to the singularity of the Earth is fighting to destroy the singularity of the top which is the movement of the Earth is fighting to destroy the movement of the top and all of these relevancies are all the same.

This issue is of cardinal importance and could deliberately be altered to hide the misinterpretation science wishes to connect to mass in order to hide the fact that mass does not bring on gravity but it is gravity that brings on mass. Mass is achieved when the object is resting motionless on the surface of the Earth while it is gravity that is still attempting to obtain movement as to try and move the object down to the centre of the Earth. This movement consists of two parts where one part is following the curve of the Earth while the Earth is rotating and the other part of the same movement is the thrusting of the object educing the object to move to the centre of the Earth. Mass is the result of gravity and not the other way around. Gravity brings on mass and mass depends on gravity to have any value or function.

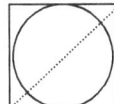 **The value of singularity stems directly from the law of Pythagoras** or **Pythagoras** is the result of **the average of singularity. With the shortest line being a dot, all lines must start from a position implicating** Π.

A circle is a square without corners implementing Π and a half circle is therefore a triangle without corners. The corners are the factor that confused every one in the past. When replacing the value we normally attach to circle being r with Π, the law of Pythagoras becomes quite meaningful and mathematical.

By   placing a connecting circle on the sides of the triangle half a circle forms. By implicating Π as a   relevancy and not the straight-line r, two values of Π applies to each circle and the straight line is no longer r, but is $Π^2$. This will bring about that each circle holds half the square value implicated to the allocated conditions applying to Π in that specific instance. By adding the two half squares forming the two half circles and then calculating the square root of the total that then forms the average diameter, an average of Π in the connecting line will come about. As both lines are the straight line forming singularity coming from one line being Π, the connecting line then must be the average of the two lines as $Π^2$. That is what **the law of Pythagoras says. Gravity is the result of the Earth spinning around its axis as well as around the axis of the Sun and the dimensional change implicates the law of Pythagoras.**

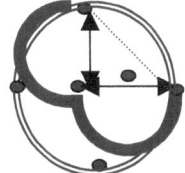 Because every moving line represents one quarter of the sphere in relation to the rest of the sphere and the line also indicate the relevant position between the point indicated and the point in the centre it is a relevancy of singularity in progress. By connecting the line, as Pythagoras will suggest the singularity within the sphere become a specific value indicated representing one half circle.

Gravity comes about as a result of the Earth turning in space and with that it pulls objects from space towards and onto the Earth. This is done by the duplication of the law of Pythagoras.

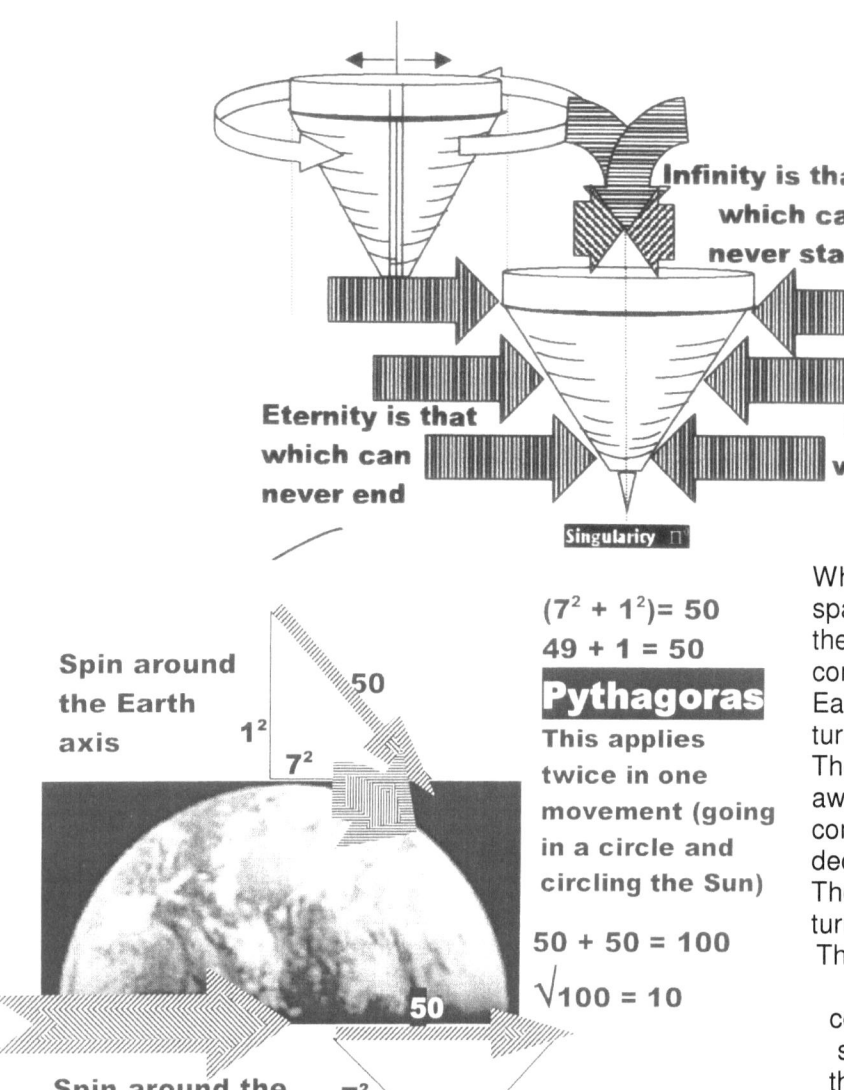

Infinity is that which can never start

Eternity is that which can never end

Eternity is that which can never end

Singularity ∩

Spin around the Earth axis

1^2

7^2

50

$(7^2 + 1^2) = 50$
$49 + 1 = 50$

Pythagoras

This applies twice in one movement (going in a circle and circling the Sun)

$50 + 50 = 100$

$\sqrt{100} = 10$

Spin around the axis of the Sun

7^2

1^2

50

When the object moves while being in space or in contact (in relevance) with the spinning Earth, the object wishes to continue moving straight ahead while the Earth also moves straight ahead by turning 7°.

Therefore, the Earth by spinning is falling away by turning 7°. That clears space or compresses space by the margin of 7° declining (compressing) of air / space.

The Earth spins around its axis by 7° and turns around the Sun by 7°.

The Earth is moving, constantly spinning and in this is contracting space by compression (we call this contracting of space in air the atmosphere) and while the air is getting more compact, it takes whatever is filling with space towards the Earth constantly at a rate of 7°. By the Earth rotating, it is compressing space and with space compressing it is moving objects in the direction of the Earth. That is why objects that is falling, has no mass and only the stupidity of the simple Newtonian mind will force scholars to accept that it is mass that is pulling gravity.

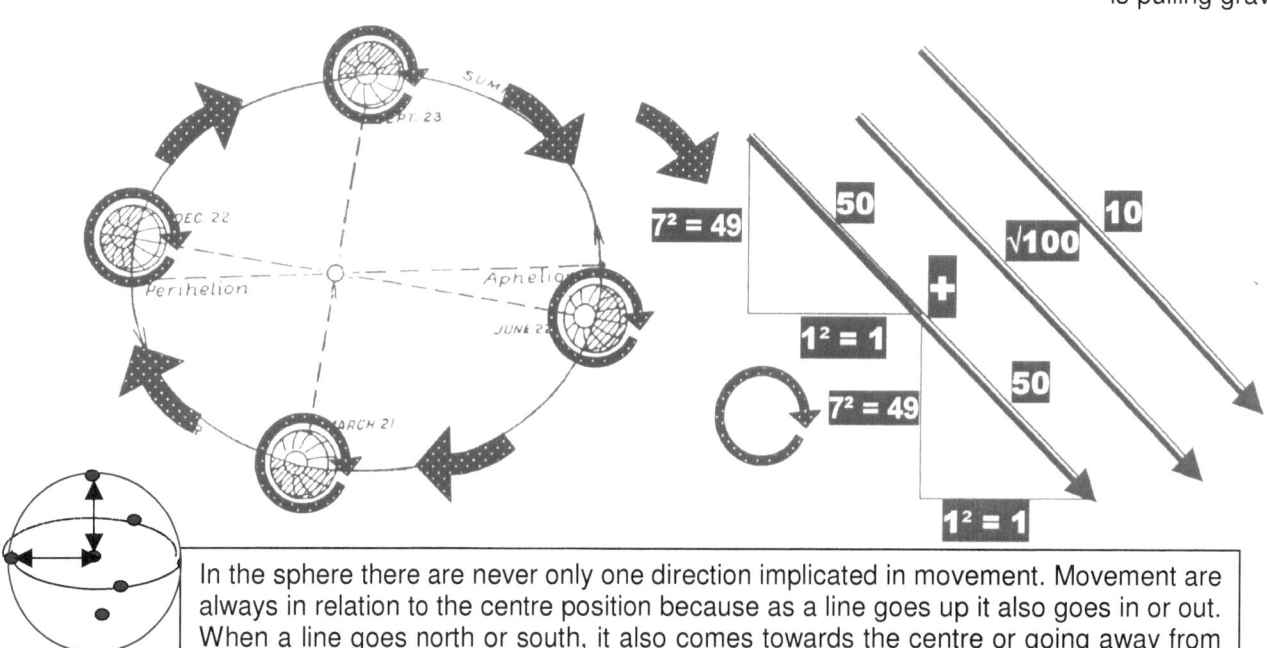

$7^2 = 49$　50　$\sqrt{100}$　10

$1^2 = 1$　+

$7^2 = 49$　50

$1^2 = 1$

In the sphere there are never only one direction implicated in movement. Movement are always in relation to the centre position because as a line goes up it also goes in or out. When a line goes north or south, it also comes towards the centre or going away from the centre.

There is always relevancy present in movement. As this moving indicates direction it also applies Π^2 for indicating value forming the time factor.

In the sphere there is no radius but only the extending of Π from the centre Π^0 going in six opposing directions relating to one another by the square but remaining Π because of the unity the matter holds in relating to space. Every opposing point unifies by the joint forming of Π. If we wish to bring this reasoning into the cosmos where atomic locations apply it is not possible to draw a precise line that would form a precise ring and not cut some atoms in parts.

Looking for gravity applying there will always be an atom disallowing the precise positioning of the circle, but with gravity the circle continues on a solid basis holding Π as a positional reference and not r. Using gravity one has to apply Πr^0 to realise the effect of gravity or Π and not r. In every sphere there then are the seven points confirming Π relating in precise dimensional and positional equality forming equilibrium to the centre Π^0 as well as to one another by 90^0 and 180^0 implicating the dimensional positioning. Therefore the sphere holds $_7 // {}^\Pi$ and the cube holds $6 \times r^2$. This argument is very important when studying the Cosmic Code.

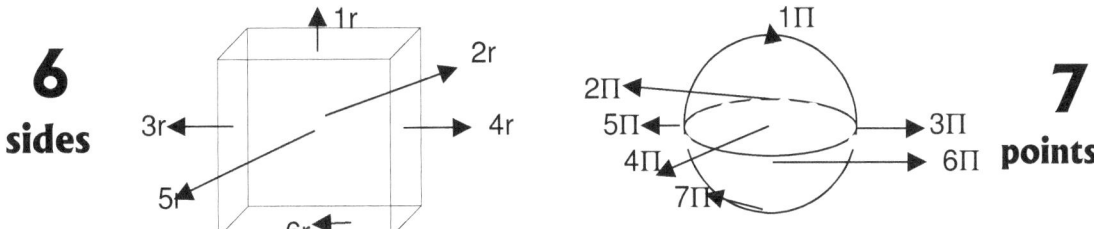

6 sides   **7 points**

Where space comes into contact with the sphere the cube loses one of the six dimensions it has to the more dominating seven dimension points forming a unit by employing Π in the case of the sphere whereby the seven dimensions in equilibrium by Π interacting will dominate the six dimensions or sides of the cube that are loosely connected by r. In this arrangement the domination of the sphere will always remove one point of the cube bringing about that the cube then has 5 sides to the seven of the cube.

This means that in the cube the "bottom falls out" and without a "bottom" to support objects they fall to Earth. Remember that a body "floats" in space, where the cube is not in contact with the Earth as a sphere but at one specific point all objects starts to "fall" to the Earth and it is at that point that the sphere comes into contact with the cube. The fact of gravity is that gravity is in place in relation to form holding reference where mass is not implicated.

5 sides in the cube vs. 7

That too is the Lagrangian system with five cosmic structures holding relevancy to the centre structure where the centre structure stands in for seven positions diverting from singularity and the orbiting structures standing in for five positions in space.

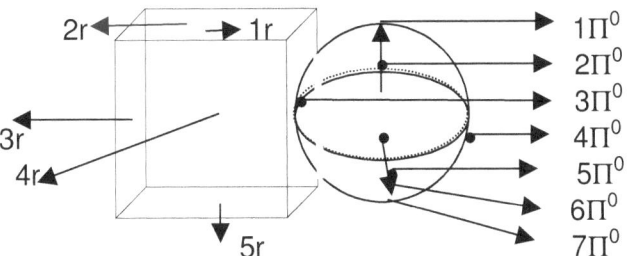

Around axis by 7^0

Around the Sun turning by 7^0

Everything in the Universe that is spinning and that includes everything anybody can think of because being part of the Universe means spinning in the Universe uses two axis holding singularity around which it spins. The one spin defines $\mathbf{T}^2$ while the other defines $\mathbf{k}$ and in that the definition is about the space $\mathbf{a}^3$ that forms. Both spins are putting 7^0 in relation with singularity and since singularity is 1^0 it puts $\mathbf{T}^2$ as 7^0 and $\mathbf{k}$ as 7^0 and that brings about $(7^2 + 1) + (7^2 + 1)$ implements the Pythagorean Theorem. This puts double seven by the square hypotenuse as a right triangle $(49 + 1) + (49 + 1) = 100$. With The Pythagorean Theorem the hypotenuse is 10 ($\sqrt{100} = 10$).

The following is not an in-depth investigation into the four cosmic pillars or the four cosmic phenomena but is a sketchy overview glancing very broadly at what forms the four phenomena. The phenomena use form and are the result of $\Pi^0\Pi$ forming space-time where the curvature of space-time results from Π manifesting as space-time. In short, it shows the truthfulness of the four phenomena resulting in gravity forming. The Titius Bode law put double 7 as being found in $(7^2 + 1) + (7^2 + 1)$ in relation to 10 as well as 10 in relation to 7.

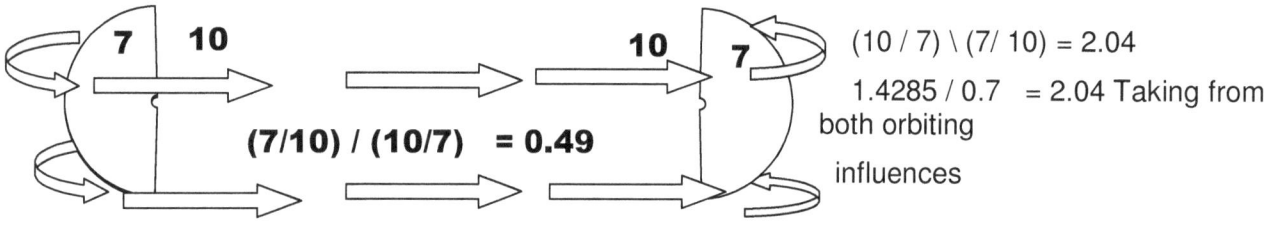

$(10 / 7) \setminus (7/ 10) = 2.04$

1.4285 / 0.7 = 2.04 Taking from both orbiting

influences

(7/10) / (10/7) = 0.49

SPACE DIVIDED INTO TIME

(7/10) / (10/7) = 0.49

.7 / 1.4285 = 0.49 Taking from both orbiting influences

THE PROCESS PARTED USING THE ROCHE PRINCIPLE

10 / 7	$(\Pi/2)^2$ The Roche influence on Titius Bode
7/10	$2.04 \times (\Pi/2)^2 =$ 5.033
$(\Pi/2)^2$	$2.04 \times (\Pi/2)^2 =$ 5.033
10 / 7	5.033 +5.033 = 10.066 from both objects

Crossing the singularity divide and activating the Roche principal $(\Pi^2/4)$

SPACE MULTIPLIED WITH TIME

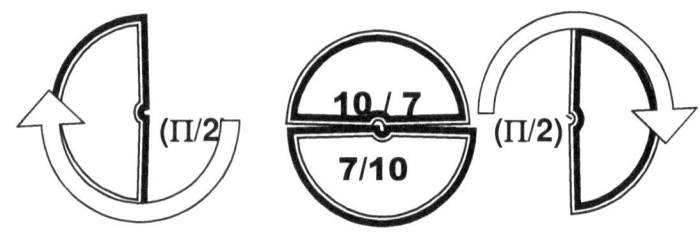

The crossing of the divide will implicate singularity on both sides of the divide bringing about the Roche factor

10 / 7 $(\Pi/2)^2$ The Roche influence on Titius Bode   7/10 $2.04 \times (\Pi/2)^2$

7/10 / 7/10 = 1 and   10 / 7 X 7/10 =1

Those factors all being equal to each other while space holds a value of 10 and material by movement holds a value of seven and therefore equal to one is not influencing change. The space that the motion establishes creates a relevancy of seven factors in space while the direction of motion involves another three dimensions or points, which is incorporating the other singularity in the unit. While the motion is at the same time moving out of ten points in relevancy and only occupying seven points the very opposite comes about through the same action being duplicated. The motion turns ten points to seven by moving from ten and filling seven points through the motion ending before the next cycle starts.

SPACE DIVIDES INTO TIME

On the one side of the Universe

7/10 /10 / 7= 0.49

on the other side of the Universe

7/10 /10 / 7= 0.49

And on both sides of the Universe

.49+.49 = .98

(10 / 7) \ (7/ 10) = 2.04

.98 X 10.066 = 9.86=Π^2

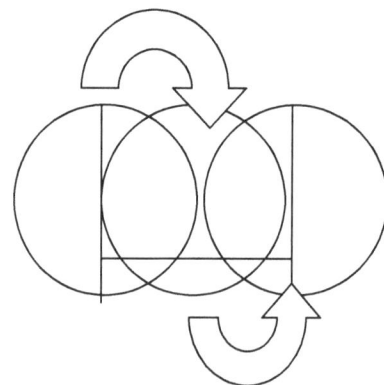

The value science use for gravity is 9.81, which they measured to much detail but the moon has a lot to do with the recorded difference coming about.

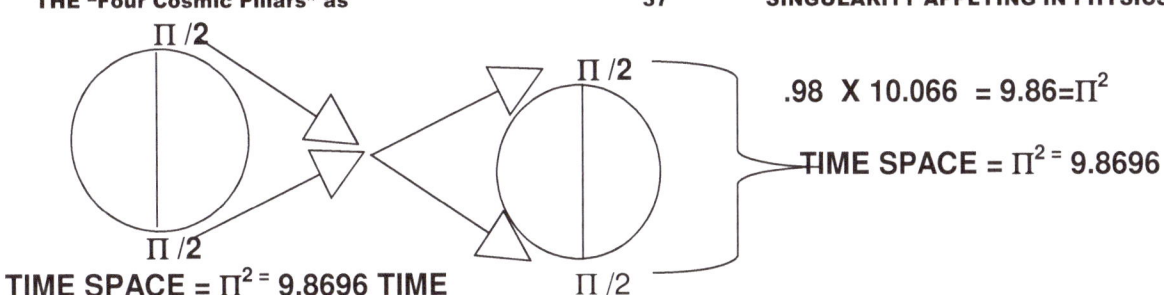

$$.98 \times 10.066 = 9.86 = \Pi^2$$

TIME SPACE $= \Pi^2 = 9.8696$

TIME SPACE $= \Pi^2 = 9.8696$ TIME

The TITIUS BODE Principle Outside the sphere

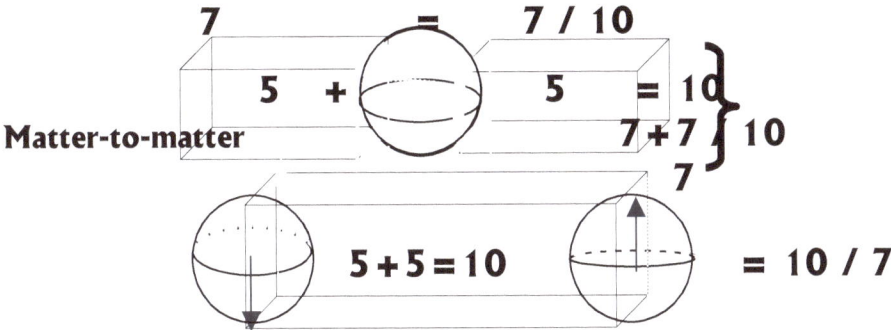

Matter-to-matter

With the dimensional change from space in the cube to space in the sphere a relation of 5 to 7 comes about depicting gravity. The principle of 5 sides in space relating to 7 in the sphere holding matter forms the basis of the Titius Bode and the Lagrangian principles.

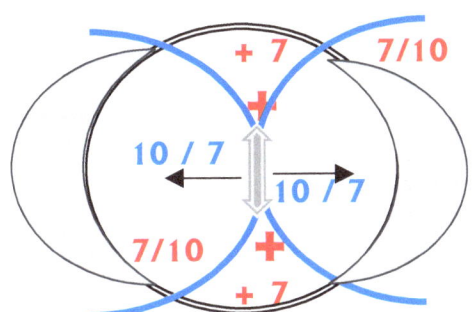

Gravity is motion in space and motion through space. Gravity is a relevancy between space travelling and the time it takes the space to travel. Gravity produces or reduces space during a certain period of synchronized spin of material in motion. Gravity is $a^3 = k\,T^2$ where it then becomes $k = a^3 / T^2$ In the light of this all other explaining fails the test of accuracy. It is no force because mass depends on gravity and gravity does not depend on mass.

The Titius Bode principle is a relation where space is the factor holding ten and material is the factor holding seven. By space being diminished by material, one relation comes about and where material dismisses space, another relation of seven to ten comes about. Gravity is the motion where space conservation is applied by motion control. By applying motion in sequence to space movement Universal harmony is installed. By applying motion at a faster pace than space conserve motion, the motion that is also gravity controls the space by motion duplicating space at an even rate. In that event gravity is applied in the manner we recognise the working of gravity and space.

The Titius Bode law is an extending dynamic deriving from the law of the gravity dimensional factor where the space factor in a square of ten relates to a matter factor in the square by half (half since nothing can be in two places in the Universe simultaneously) of the matter factor of $\Pi^{0(7+7)}$ or the square of space (10) relate to the matter factor of 7. From such a point every other point will be opposing any other point not pointing in the direction to which the first point is pointing, whereby it extends the direction it holds. No matter what the point is or where the point leads, such a point holding a specific direction will be unique in the direction it is rotating because at that or any other specific point wherever, it will be directing not in the direction it spins but in the direction flowing from the centre point outwards.

The explanation concerning the four cosmic pillars is rooted in the understanding that cosmic growth lies at the heart of relevancy transfer of time or heat from singularity the liquid to the solid or in other words from space to the atom. The Hubble growth is what life would see as healing and aging or "the moon going further away from the Earth" or "the Earth growing bigger around its circumference". It is the atom that is presenting the solid that exceeds its size to the determent of liquid. I explain this process in other books such as **The Veracity of Gravity** and **an Open Letter on Gravity.**

The solar system grows at the heart of time and that is where space becomes three dimensional as it does where Π becomes $7/10\Pi^6/6 = 112.16$ and this code I explain in the **Cosmic Code**. The cosmos grows at Π before it expands by movement to $7/10\Pi^6/6 = 112.16$. The actual process of expanding and the way it involves Π is far to elaborate to explain in a short introduction book such as this.

Gravity is in place when space is in motion producing a duplication of lesser space than time will form as an ongoing sequence. In the other scenario overheating or space expansion or antigravity will come about when gravity cannot sustain space duplication in the preventing of duplicating to bring about reduction. Gravity I the movement of time in which space around material reduces due to the movement of such space filled with material.

With space holding material and remaining at an even duplication without adding heat such space not holding material will become lesser dense and such thinning or loss in density will also increase space as antigravity, which is the same thing as overheating, and space growth comes about. The duplication that motion provides produces cooling in the material sector and this explaining requires far too elaborate detail and that is left to books I have already mentioned.

The duplication that motion secures prevents or supports space in motion by forming harmony in frequency to the motion, which is in truth duplicating space. One side of space is the duplicating of the other side because time is eternal at singularity. It is the double motion applying that performs the gravity between the objects in rotation. But the way Kepler introduced it diverts drastically from the way Newton introduced gravity. The moving away in relation to the coming towards us has when motion applies forms the square that we see as the rotation. The rotation does not bring about an accomplishment of zero as Newton suggested but it brings about a square, which Kepler introduced.

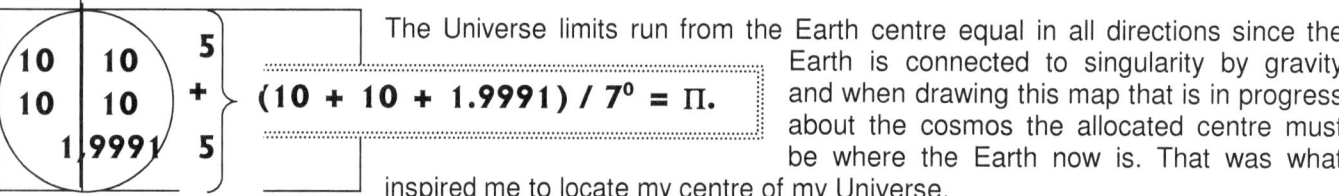

The Universe limits run from the Earth centre equal in all directions since the Earth is connected to singularity by gravity and when drawing this map that is in progress about the cosmos the allocated centre must be where the Earth now is. That was what inspired me to locate my centre of my Universe.

Even admitting to such a notion sounds like madness, but allow me please the opportunity to explain in more detail. I realised that my effort to locate the point holding singularity enabled me to backtrack the exploding Universe to its origins. By applying some basic effort I have located the position from where all movement came and the direction it took moving forward in time...and yes, even time as such. Gravity is the dimensional changing of space holding r as reference in the cube as to the sphere holding Π as the reference. In order to generate spin that is producing time in matter occupying space, therefore creating dimensional change, Π has to be a factor indicating the possibility of spin because by implementing Π the circle sides will follow one another without establishing separation.

As soon as motion takes gravity straight, singularity will reposition the direction changing the direction of motion by 7^0. It is this turning of motion by redirecting the continuing of motion that sets the critical time within the proton connecting to singularity. Instead of r being a line, gravity will inevitably be Π, which is the form value of singularity. That is this 7^0 redirecting in the square of space, which is ten on both sides of singularity and time is that what we find to be the Titius Bode law of $7/10$ and $10/7$ in relation to the Roche limit of $\Pi^2/4$ which is producing the gravity of Π^2.

However the reducing in it is going from ten that is on one side and is crossing over the figure of 1.9991, (which is singularity on both sides of the Universe) and coming into contact with another 10 while turning 7^0 that we find to form Π. In all being the total forming on both sides of the Universe it is $(10 + 10 + 1.9991)/7^0 = \Pi$. The answer must be in finding Π, and thereby locating singularity. If singularity is in affect the original point of the cosmos birth, the reducing path we should follow will indicate the whereabouts such a point must be. That is where cosmology diverts from mathematics.

The seven can never totally separate from the ten, but by singularity being the same but being on the other side it is withdrawing space-time altogether. See it as seven (let us think of that as the cold basis of space) spinning or turning in the ten (which then will represent the hot part in the cold basis) and the ten is part of the seven but the seven is not part of the ten.

The third factor is the axis around which hot as well as cold will turn. Therefore when reading the next page please envisage a cold base turning in a hot and cold space. The purpose of this is not to define whether the argument is correct or not but it is to help **the reader** gain understanding of the process principles involved. But motion also converts space to relate to space by changing relevancies through motion matter is in relation (part of) to the total dimension of space but is not the total dimension of space.

This is the prime element the state where everything started. It started at the time when mathematics was still to be invented by nature and only singularity was in place forming a value of Π and a reference of 180^0 in all directions in relations to other positions singularity established. It is at this point where everything other than **singularity was $\Pi^0\Pi$ becoming Π going on to be Π^2 through motion forming Π^3.**

One must take into account that gravity is different motion of particles forming a relevancy about duplicating space and dismissing space in relation to the effort of the particular and specific element. The motion differences in the motion between two particles bring about relevancies. It is a seven factor standing in relation in motion to a ten factor of which the seven then is included and part of the ten factor. The four time factors that I refer to as the four pillars are applying gravity as time is on the edge of the sphere in relation to the centre line forming singularity in the sphere.

Science (as usual) does no understand the true complexness in the phenomenon applying and then try to substitute ignorance wit fact they too do not understand. In science they mistakenly refer to the Titius Bode effect as the Doppler effect because Doppler found the sound of a train moves in circles which then forma a relation to the sound moving forwards and the sound moving to the back. The Doppler effect is the Titius Bode law and the sound that Doppler discovered works by principle on the very principal that is manifested in the configuration we witness as the Doppler effect. That which Doppler saw the day with the steam train producing sound waves was gravity carrying the sound in wavelengths and this method is a manifestation of the Titius Bode law. It uses the Titius Bode law but the movement of the train distorts the wave pattern of 10 / 7 somewhat.

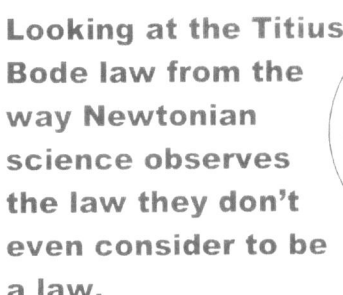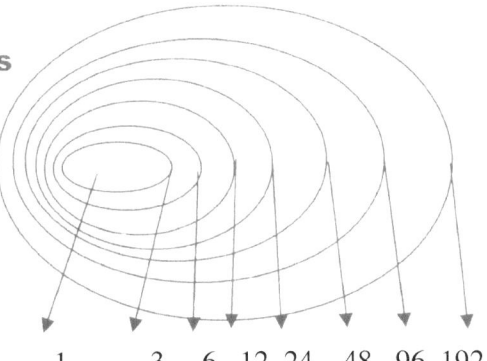

Looking at the Titius Bode law from the way Newtonian science observes the law they don't even consider to be a law.

1   3   6 12 24   48 96 192

Planet	Mercury	Venus	Earth	Mars	Ceres	Jupiter	Saturn	Uranus
Bode's Law distance	4	7	10	16	28	52	100	196
Actual distance	3.9	7.2	10	15.2	28	52	95	192

Bode's Law:
A numerical sequence announced by J.E. Bode in 1772, which matches the distances from the Sun of the six planets then known. It is also known as the Titus-Bode law, as it was first pointed out by the German mathematician Johann Daniel Titius (1729-96) in 1766. It is formed from the sequence 0,3,6,12,24,48,96, and 192 by adding 4 to each number. The planets were seen to fit this sequence quite well – as did Uranus, discovered in 1781. However, Neptune and Pluto do not conform to the 'law'. Bode's Law stimulated the search for a planet orbiting between Mars and Jupiter that led to the discovery of the first asteroids. It is often said that the law has no theoretical basis, but it does show how orbital resonance can lead to commensurability. The importance that becomes known is the sequence the Titius – Bode law saw in the number arrangement of 3; 6; 12; 24; 48; 96 etc. The incorrect application of the Titius Bode law lies in subtracting the figure of 3 from 10 leaving 7. The other way of reasoning is to add four each time to the firs value of three starting with 3 and so on. The true significance of the Titus-Bode law is that it points directly to a circular growth of 7 stages. The 7 relating to 10 is a precise derogative of the Roche limit or the Roche limit is a precise derogative of the Titius Bode principle because he two systems interlink.

Whenever any circle comes about the value of Π becomes an issue not to be ignored. In concerning Π it is most important to look at the factors forming Π and I am not going into that argument very intensely as I try to keep as simple as possible in this book where I introduce the concepts. Looking at gravity is implicating Π and therefore the factors forming Π have to bring about gravity. Notwithstanding any wizards' ideas about mass forming gravity, gravity is about maintaining circles and movement. That is maintaining Π.

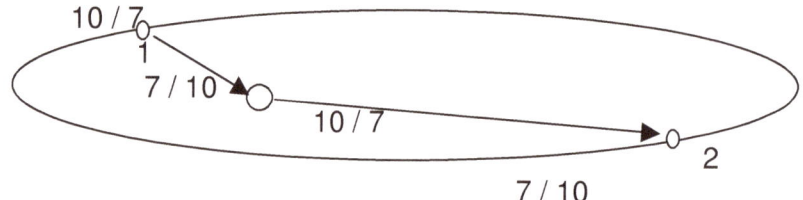

When a circle forms it implements Π as a form and therefore the factors forming Π has to charge the circle forming by the use of Π. In the value of Π we find 7 in relation to 10 and we have 10 relating to seven forming at the same token a value of ((7+7) and 10 / 7), which converts to $Π^2$. Whenever anything orbits another thing gravity comes as such a result and gravity comes about when 7 and 10 interacts. I have already shown that gravity is double 7 interacting with singularity and why that is the case. I have also shown how 10 come as a result from this by implementing the law of Pythagoras.

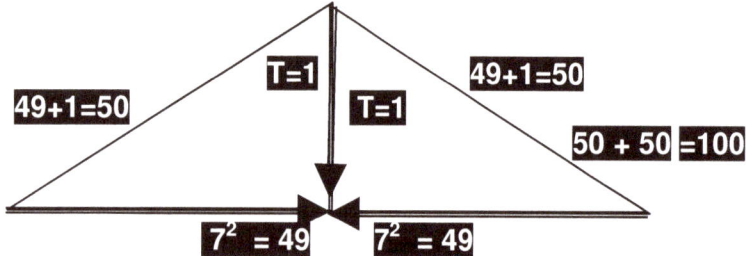

By establishing gravity we find the conduct of singularity interacting and therefore establishing the value of Π. However at the point Π is validated the Universe is still flat.

The explaining of the following is a little bit extraordinary but so the understanding of any of the factors forming cosmology. By using singularity the use of singularity not only involves the equal measure of singularity but also in principle being singularity means there can't be two being equal. Being singularity means there is one point holding singularity where it is the same singularity because space is parting singularity is multiplying the very same singularity since singularity is 1^0. Therefore space can be multiplied but not as it involves singularity.

Multiplying one with one result in achieving one 1 x 1 = 1 and therefore one is not just equal to one but is the very same one 1 = 1 notwithstanding whatever space forms between the points representing the same singularity because the space forming is also singularity. By multiplying one with one result in achieving one 1 x 1 = 1 becomes the repeat of one (1 = 1) and not the duplication of one (1 x 1). Therefore by becoming Π the cosmos is still flat. It is when a flat Π interacts with another flat Π in relation to having the same singularity $Π^0$ that we find gravity $Π^2$. It is time taking the Universe to a single dimension while space will always be 3 dimensional.

On the side forming space we have 10 being the contribution space makes forming a double as the one planet is responsible for forming 10 and the other planet is also responsible for forming 10 as part of their contribution to space while the third structure stands still ($Π^0$) as to give time validity to form while time shows growth coming from 0.991 to 1 as the factor forming Π. This is space that establishes Π. This then stands in relation to the 7^0 of spin that forms the Earth's (or any other planet for that matter) gravity.

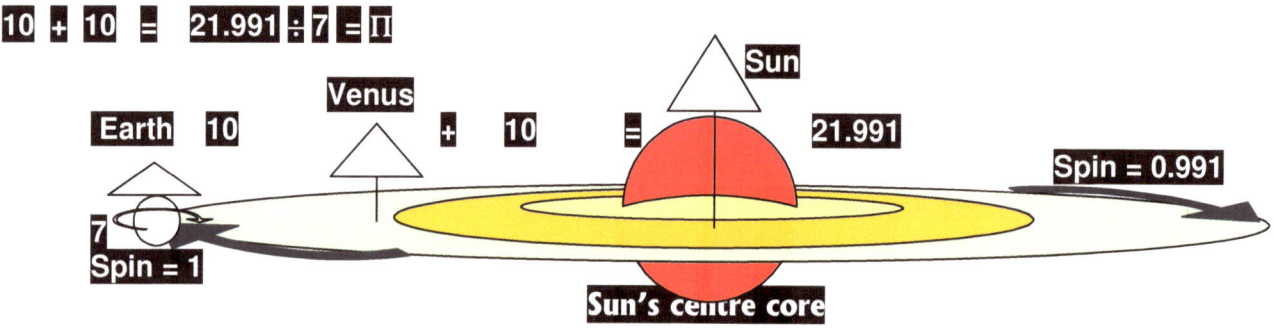

Then on the material side we have three (two planets and the Sun) all spinning which is redirecting movement to the tune of 7° and 7+7+7+0.991= 21.991. This again and once more stands in relation to the 7° of spin that forms the Earth's (or any other planet for that matter) gravity in relation to the atoms within the Earth.

By having space forming Π and material spinning forming $\Pi^o\Pi$ the interaction of space-time then will become gravity Π^2. That is the one way that the relevancy of space-time takes place forming 3 dimensions. It must again be said that time $\Pi^o\Pi$ goes flat while space is moving $\Pi\Pi^2$ remains 3- dimensional.

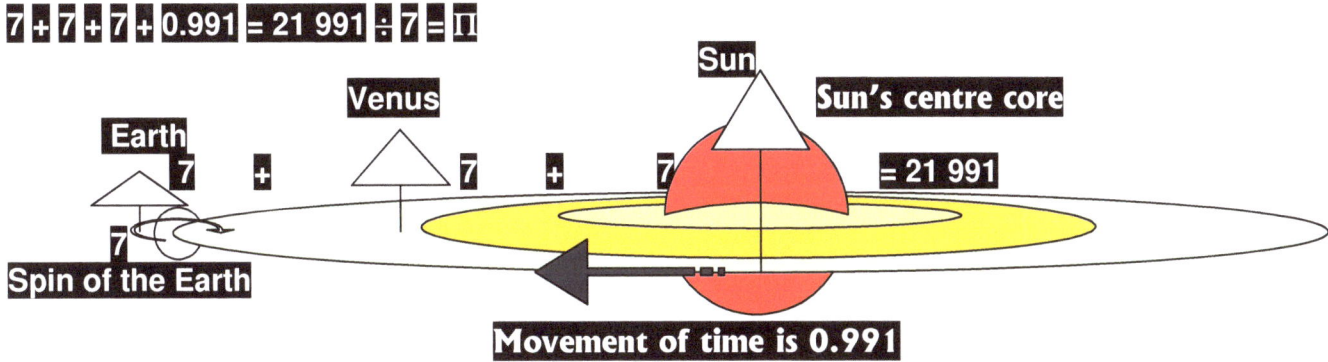

With every planet spinning in synchronised order the three forming the Titius Bode law will each contribute 7° of spin that adds to Π forming. In relation to that we have on the side forming space a double 10 forming in relation to 1 standing still and the movement coming about contributes the growth time contributes that is 0.991.

With material forming Π (7+7+7+0.991=21.991) and space contributing (10+10+1+0.991=Π) we fins time forming Πx Π=Π^2. The movement we see forming gravity is not in the one but is in seven duplicating by the square to form 49. The one can never move because the one is singularity and even if singularity goes square (move) it remains 1.

Looking at the overall picture we may find that 7 spins around 1 but since 7 spins around the same one it is spinning the same 7 that then creates a dual in 10 where ten serves as a value that forms space. The mathematical implication in the Titius Bode law works on the principle that we find the number seven doubling as every planet holds a governing singularity that works in tandem with a controlling singularity and the direction changes of the two factors always work in tandem.

This is most impotent.

One can never see one singularity bring motion about without the other principle also influencing the outcome. From every planet the value of the 7 applying will change and in every case another double seven will be in place since every planet is a Universe apart from the rest of the Universe . In the picture forming space-time by explaining the Titius Bode planet layout we have 7 by the double as well as the planet holding its own position according to the Sun which is 10.

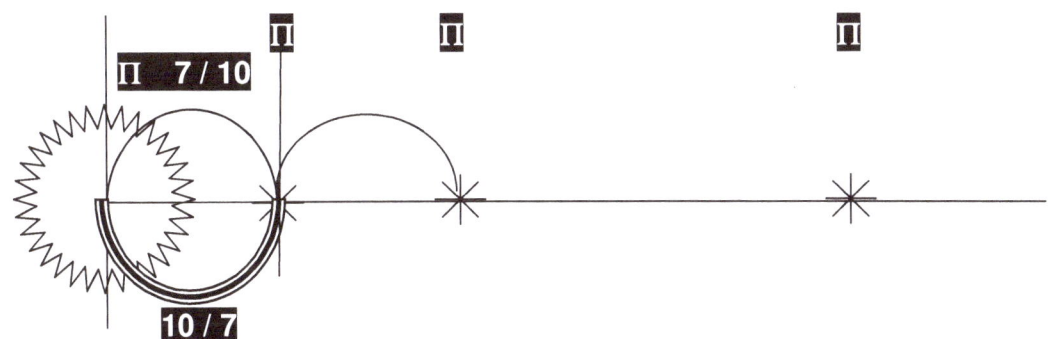

The reason why Mercury has such an "abnormal" orbiting route will be the absence of a double 7 guide.

Any changers occurring in Π will lead to a an unequal triangle providing two different values to r and will alternate the link between r and Π^2 bringing about different form (Π) and time (Π^2).

When singularity forming the lines of the triangle is not in equilibrium the triangle will destroy the matching of half circle.

The spherical positioning layout forming the Titius Bode Principle

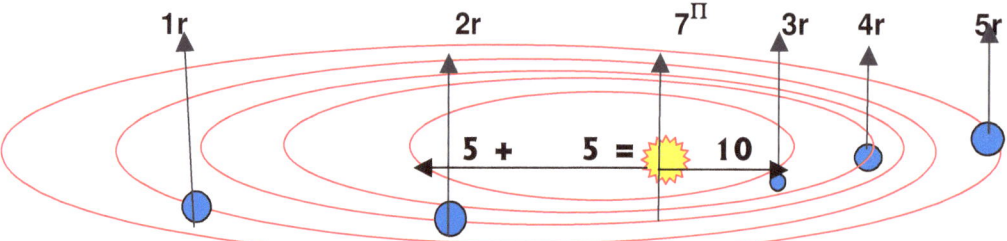

From the matter-to-matter relation in the Titius Bode configuration there are 7 / 10 + 7 / 10 = .7 + .7 = 1.4

From the space-to-matter relation in the Titius Bode configuration there is 10 / 7 =1.42
The 5+5=10 is a position of dimensions as space loses value to singularity. The 7 that matter diverts in points from singularity may seem as coincidental but is valid. Still in accordance to our perception valuing the number in degrees, it seems coincidental but if it is coincidental, it is nevertheless a figure of diverting proven as accountable in all other calculations and plays a most dynamic role.

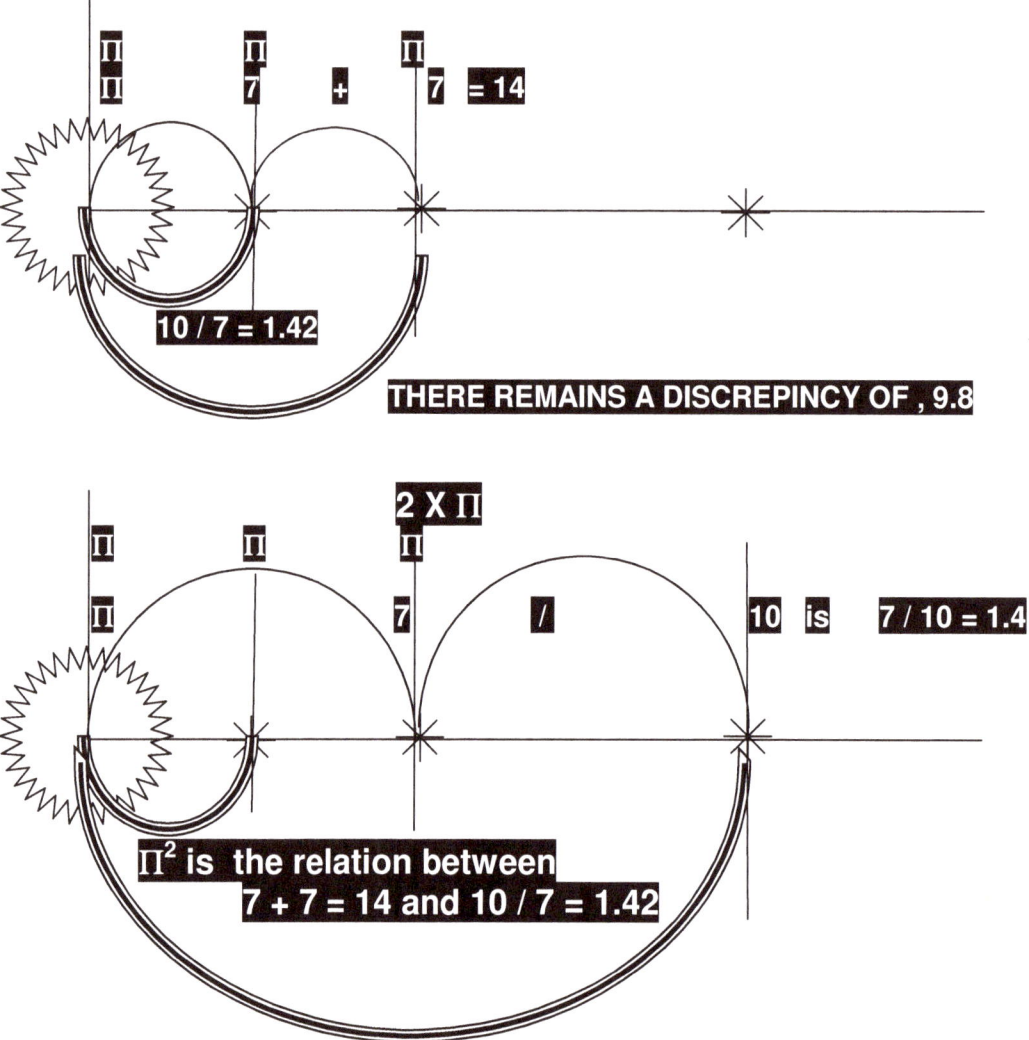

The normal application of the Titius Bode Law provides an interaction between the orbiting object going into gravity by spinning in a double seven synchronized manner and into y then forming a compiling value of 10 which is the square of a hundred which is the double fifty which is the result the hypotenuse is the law of cosines which is mathematically $a^2 + b^2 = c^2$ or $(7^2 + 1^2) + (7^2 + 1^2) = 10^2$. It is the manner in which time builds space by spinning those implementsΠ. It is Π forming space –time by (10 + 10 + singularity 0.991 expanding to 1 being in relation to gravity @ 7) It is Π developing space-time.

In order to try and simplify the explaining there is anther way of expressing the Titius Bode principle and that might be as follows:

Let's say the Earth will be at a larger diameter horizontally (10/7 = 1.42) than at the vertical diameter (7+7)/10 = 1.4. This proposal is that the atmosphere (space-time) at the Arctic is at a value of 7 to 10 and at the equator it is at a value of 10 to 7. The equator holds a much higher relation to heat unoccupied than does the Arctic region and this will allow the atmosphere to contain much less density than the density applying at the equator. This will favour a space value of 10 in the equator regions while the double 7 will represent the governing singularity line running from top to bottom. This puts the line in relation to the circle forming by spin.

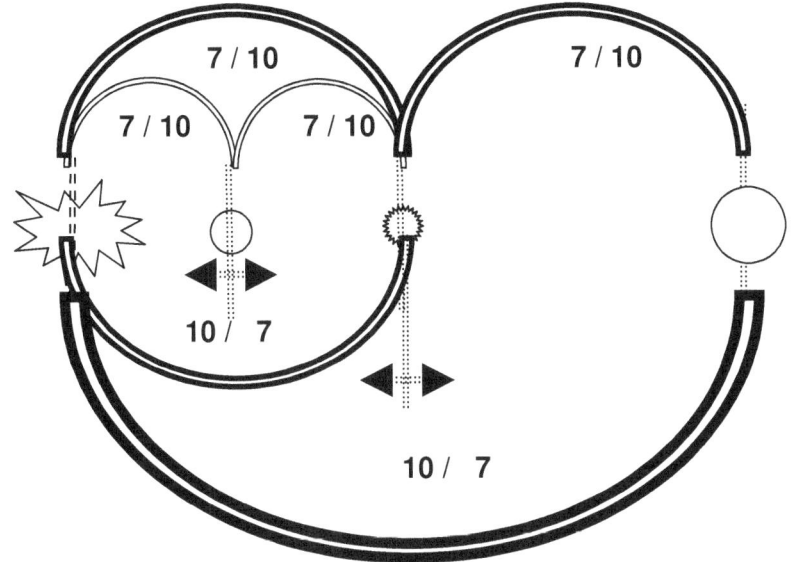

The Titius bode Law is holding 7/10 and 10 / 7 as it extends its influence much wider, but the influence of the extending as far as the Earth is concerned is merely a drop in a bucket to the effect it holds on gravity through the Sun's influence. That is how the Universe's expanding works. Little by little time moves onto become space at the point where singularity introduces time that then becomes space. One must acknowledge that space is the history of time and time leaves space as a reminder of time that has gone by. This is how space-time builds and that is why we detect this as the building form we see in the solar system.

To explain the Titius bode principle in detail one must once again return to singularity and understanding that is as simple as it is totally uncomprehending.

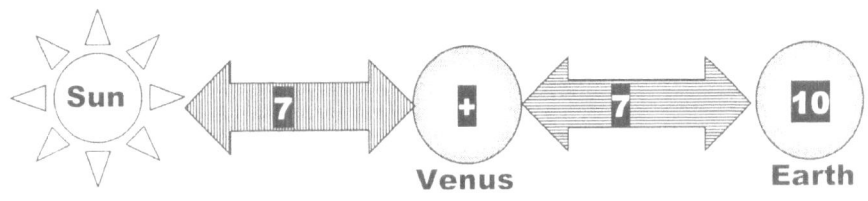

Now we look at the Titius Bode law from the way the cosmos applies the law

Fact 1: The importance that becomes known is the sequence the Titius – Bode law saw in the number arrangement of 3; 6; 12; 24; 48; 96 etc. This goes beyond doubt. That puts Venus at 7 and that puts the distance between the Earth and Venus also at 7 because the distance always doubles according to the Titius Bode law

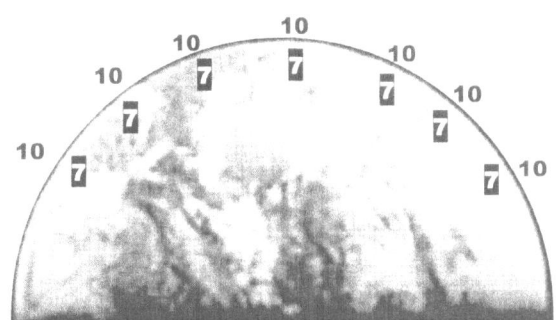

is = $(\Pi/ 2)^2$ = **2.4674.**

In this we find the value of space-time compiling. We have 7 as a distance from the Sun to Venus and 7 doubling the distance from the Earth to Venus while the Earth forms the time value in space of 10.

The space between the spheres divide in half, but because of the extending of Π and not applying r as ordinary mathematics will suggest where Π replaces r the singularity extending from Π^0 will be half of Π in the square of Π and that

As I am about to show mathematically how 7 relating to 10 by the same action where 10 relates to 7, this relation forms Π^2 this happens in a double spin. It is an atom that forms the star. The atom spins around its axis. That is one Π^2. Also the atom spins around the star's livers another Π^2.

The lines converging from singularity holds a square to one another and that implicates the oldest mathematical principle that I know of, the law of Pythagoras.

Again we find the presence of a triangle holding a square. This holds space away from matter and therefore we are calculating the square of space depicting singularity and time (always in a square) away from the immediate claim on space by matter.

Matter in relation (part of) with the total dimension of space.

$$\left(\frac{10}{7} \div \frac{7}{10}\right) = 2.04$$

$$\frac{1.4285}{0.7} = 2.04 \quad \text{Taking from both orbiting influences}$$

SPACE DIVIDED INTO TIME

$$\left(\frac{7}{10}\right) \div \left(\frac{10}{7}\right) = 0.49$$

$$\frac{0.7}{1.4285} = 0.49 \quad \text{Taking from both orbiting influences}$$

SPACE MULTIPLIED WITH TIME

$$\frac{7}{10} \div \frac{7}{10} = 1 \quad \text{and} \quad \frac{10}{7} \times \frac{7}{10} = 1 \quad \text{Therefore not influencing change}$$

THE PROCESS PARTED USING THE ROCHE PRINCIPLE

$$\frac{10}{7}$$

$$\frac{7}{10}$$

$$\left(\frac{\Pi}{2}\right)^2$$

$$\frac{10}{7}$$

SPACE DIVIDED INTO TIME

$$\frac{7}{10}$$

$$\frac{\dfrac{10}{7}}{}$$

$$\left(\frac{10}{7} \div \frac{7}{10}\right) = .49 \qquad \left(\frac{10}{7} \div \frac{7}{10}\right) = .49$$

$$\left(\frac{\Pi}{2}\right)^2 \text{The Roche influence on Titius Bode}$$

$$2.04 \times \left(\frac{\Pi}{2}\right)^2 = 5.033$$

$$2.04 \times \left(\frac{\Pi}{2}\right)^2 = 5.033$$

$$5.033 + 5.033 = 10.066 \quad \text{from both objects}$$

$$\left(\frac{7}{10}\right) \div \left(\frac{10}{7}\right) = 0.49$$

$$.49 \quad + \quad .49 = .98$$
$$.98 \times 10.066 = 0.8 = \Pi^2$$

$$\text{TIME SPACE} = \Pi^2 = 9.8696$$

TIME SPACE = Π^2 = 9.8696 = Space and time in a dimensional implication

A STRAIGHT LINE, TRIANGLE AND HALF A CIRCLE WILL ALWAYS HAVE EQUALITY IN DIMENSIONAL CAPACITY PROVIDING EQUILBRIUM BEING 180^0 BECAUSE EACH ONE SHARES A COMMON DINOMINATOR IN SINGULARITY TO THE VALUE OF Π. As the straight line averts a zero it holds another straight line in place to set about such an averting where the two lines will always carry a relevancy in elation to progress (the triangle) and a common denominator in the start from singularity. This concept we apply as the graph or the vector.

 When decreasing the radius to a point where such decreasing will bring about going across the circle centre to the very other side of the centre of the circle and while doing so would bring about that this will lead to the entering a zone with no space where entering this area would consuetude in immediate landing on the opposing side where the spin then will be in an opposing direction. At such a point where there is no further space the decrease of the radius will no longer affect the value of pi. It is only from a straight line r growing that such growth will influence the appreciation of Π, but to influence Π would lead to a increase in r as Π and r are different entities. Therefore when only Π is left and r^0 has gone singular then no space is left because Π only holds form and only form is what then is left. Looking at the affect of gravity it shows the precise quality of no distinctive point, as gravity never seems to end at a point but flows all over affecting all that holds a position in its sphere of influence. The gravity coming from China meets the gravity coming from America at no particular spot but intermingles without distinction.

By tracing the line back to where the circle is no more, the reducing of the straight-line will uncover singularity plus one dimension valuing Π. But while the entire centre forming singularity is still locatable within the Universe we have it is not holding any dimension we may recognise. Reducing the radius r from all angles possible throughout the circle will bring about that all possible direction will eventually land on the very same spot with no more dividing possible. Yet zero cannot be a factor since the sides still hold value, in as much as holding all the value there can rise from such a spot. This is arriving at a point where more reducing will land the one side on the opposite side of the line but it will still not equate to zero.

What this argument further proves is that the circle reducing must then come from all points because the radius might be a line but that line represents a circle through 360^0 coming from and accounting for all possible directions. Taking that into account, it is important to recognise that notwithstanding the size of a line, which any radius of any size is, there is another line (or dot) eternally bigger as well as eternally smaller than the line in question. While we are in the third dimension being part of the third dimension such being in the third dimension then allows that all parts of the third dimension forever can be divided once more until the line in the third dimension is no longer part of the third dimension. When such a line leaves the third dimension it is still dividable because it might not be part of our dimension any more but it can still reduce further as part of the second dimension. By moving from Π^0 to Π, such movement constitutes to forming the second dimension by forming a square of Π or Π^2. By that time it has left our scope by miles but that does not mean that it ends there because from our perspective that is where it ends. But our perspective does not represent reality. Yet, even then it can still reduce infinitely more from Π to Π^2. By such reduction it forms part of the second dimension and then at last when going Π^0 it forms part of the first dimension. The Universe is Π by never being Π because it will be Π^0 and by moving from Π^0 to Π the movement will bring about Π^2.

However in reducing when the line reaches the first dimension at that point no further dividing of that line is longer possible. We can never grasp what the size of a line that first line is in size that comes about when the first motion breaks the eternal stranglehold of singularity on space. According to our big and small conceptions of what we perceive as large, ultra large, small and microscopic small is just mere words describing thoughts that is totally unrealistic in the context of what the cosmos sprang from as the cosmos moved out of the spot and formed a dot. Even by the standards of forming the dot, which was eternally bigger that the spot, from that the dot developed and developed all the many dots that came from the spot. The size differentiation that is in place when compared only between those two points exceeds all limits we are able to fathom and divides more than what we wish to create that forms borders that we can appreciate.

When looking at the circle in the conventional manner, we persist with errors brought about in culture and not by applying some significant modern logic. Take a circle and reduce such a circle constantly to where it no longer can reduce. Reduce it to a point where only form remains part of the circle because the radius has gone beyond human measure and becomes so small it is not noticeable with what ever measuring tools man may use, then what remains is pi since pi does not indicate size but indicate form, and form is all that then will remain. In any circle or sphere the size only depend on the fluctuation of r, as a component to the circle or sphere but that does not affect the form by indication of Π in any way there may be. The conclusion I drew from following this process is that from this no line can start at zero because that will be a mathematical impossibility since no line can ever reduce to zero. A line will forever be able to reduce further becoming smaller but it can never reach zero because zero is not part of the scale on which we measure lines. If a line cannot reduce to zero it then cannot start at zero.

A line or spot starting at zero would therefore be shorter than the shortest line possible. For obvious reasons can no line, or any line grow or extend from zero because such a line must then quit zero and become something, thus abandon its original value. That would mean the start of the line has a different value to the end and a line holds conformity through out. When any line is starting from point zero and it uses the factor zero, then it can never leave zero because of the influence of being zero disqualifies any possibility of growth. But when coming from singularity π^0 and the line then had to grow in all directions at the same pace the line must then become a circle π or being three-dimensional, then form a multi circle become 3-dimensional π^3 which is the circle form we named a sphere. Since the Universe is about circles and lines connecting circles I came to conclude that flowing from this fact is that in the Universe there can be no zero improvising as a filling ingredient for the space of a point or be unfilled space. Zero is no valid factor in the Universe. In the case of the growing sphere the value of the circle is Π^0 going Π, and that is where creation must have started. That gave me the clue where to start looking for singularity. One would find singularity in the value Π and the value Π will be in all things rotating in a circle but by measure one dimension smaller. As usual I am again shooting the gun before the hunt started. Lines in mathematics do not start from zero and that is no discovery on my part that was a realisation I came too. The Universe is all about lines and the manner that Kepler pointed to the increasing of the lines by $k = a^3 / T^2$ proves growth in the composition of all lines.

To find validity in my argument one must draw this statement of motion back to the point where singularity is getting sides or said mathematically Π^0 going Π. When there is singularity there can be no sides. The one forming singularity Π^0 by measure fills no space while form Π develops Π^2 into space. The space that even the dot fills being Πr^0 does not really exist in the manner we humans see space to exist. It is a spot that is there without being there. It does not visually exist because it is not filling any substance and it cannot be recognised since it is not three-dimensional. The spot and the dot have no dimensional worth of any measure.

It is the point within the Universe I have named as **Infinity** where nothing can go smaller and anything within that point can never reduce. That point is where the entirety called the Universe begins and where everything holding substance begins.

Once one accepts the fact of singularity being present in that location, that accepting of singularity then is contradicting all the things we know and we can measure and we recognise that point being present by merit of the fact that the point referred too is not being formed by any of the things we can recognise. It is made up of everything we don't know and constitutes of everything we are unable to recognise or visualise.

In that spot there is no space. That spot holds **Infinity.**

In that space there can be no motion because there can be no space to have the motion within. It is formed as a line that is so small that our human reality by perception declare that point as not being there and the only reason why we know it is there is because of the results it left as an imprint of its not being there. We cannot detect it but notwithstanding our failure to note it we can recognise the dot on the merits of its absence and while in our Universe it is always absent, reality disallows the dot ever to be absent, because it is never absent. It cannot be absent. It cannot go absent but it can never be there where it should be in a place from where the third dimension forms and it is always present if I wish to locate it. It is infinity that can never go away.

The centre spot to which I refer we cannot see and that what it truly is we cannot detect but we know what there is has no sides to any side and has no space that it fills because what it fills is filled all presenting singularity. What there is, is not valid in our domain and yet, still we know what is there notwithstanding the fact that we are unable to witness what is valid where singularity is for what there is we cannot detect with vision but we can observe what there is with intellect. The only way such a spot can fill space is by doubling whatever it fills to become more than one in the place it has to fill. But the very instant that happens it halves the space it fills because it then cuts the space it has into two parts. On this principle all movement throughout the Universe rests. From this derives motion and nothing in the Universe is without motion because everything moves in terms of all else filling the Universe. That point instigates the Universe to form by the movement of the Universe in terms of that point's inability to move. Any motion from such a point in singularity forms the entire Universe by putting everything forming the Universe in sides we do recognise. Anything within the Universe is in one side or another side or just the other side of the Universe because from that point we have dimensions forming by movement. From the smallest ever possible dot will grow a line in every imaginable direction relating to a prospect of Π because only Π will not favour one specific direction and that puts all directions at equilibrium meaning that any form of what ever might develop from such a spot will have the end and the start being in the same position, which will also have to be a sphere as the flow outward will be equal in all directions. From the smallest spot in singularity comes a sphere.

Please think clearly, is that not precisely the commitment we find in gravity, where gravity is flowing from singularity outwards but never favouring any side? We could never explain where the gravity in China meets the Gravity in America. The nature of gravity is to never end and never begin always flowing without favouring any position applying and where it seems to favour, there is a valid explaining concerning singularity. This reasoning prompted me to look for singularity in such a spot because if the prime spot from which all came was a spot holding all, then the spot must hold the shortest line but more prominent it will hold the smallest form including the smallest circle or for that matter the smallest sphere.

That leaves the door wide open for the advancing of any radius in all possible directions. With gravity always being in the centre of a sphere where the space is least available in the entire structure (there is not even space left to fill) one finds a flow of gravity from that centre spot outwards in all possible direction even-handedly. The flow outwards is a flow inwards that concentrates space where no space can ever be. The fact that the original gravity will begin as a circle or will be a circle is the direction it will take when being the first spot created. All progress will be evenly in all direction because no direction will stand out or be in favour above any other direction at first. Moreover, what this information brings home is that through motion and only through movement does space develop in terms of a relevancy dividing singularity. I am about to introduce the second form of singularity.

Kepler said that the space a^3 is equal to the motion T^2 of the space a^3 distant from a specific centre k. That then is $a^3 = T^2 k$. Within the circle $k^0 = a^3 / (T^2 k)$ the centre holding singularity also holds gravity which is centred in the precise middle of the circle. By using mathematics in the way Kepler used it, those rules and laws used correctly in the investigating of the formula that Kepler introduced must form the basis of cosmology. Also such intense investigation then must be without Newton interfering and telling Kepler what he (Kepler) should have found and subsequently Newton's incorrectly correcting Kepler whereas instead Newton should have been looking at what he (Kepler) found because only then he (Newton) could have seen what gravity is. He (Kepler) said that the cosmos said that gravity is $a^3 = T^2 k$. The space is held in check by motion from a centre and that is the way gravity develops. It becomes more than clear that space a^3 is time by dimension T^2 and time is space a^3 without dimension k Gravity is a^3 / k but k is an addition of motion T^2.

Reading this mathematically encrypted coded formula of the cosmos given to Kepler and keeping it removed from Newton it reads as being that the space a^3 is equal to = the motion T^2 of the space a^3 in ratio to a centre k^0, which is relevant to the positioning of k. If we bring in the full equation it will be $k^0 = a^3 \div (T^2 k)$ which means half of space is solid $k = a^3 \div T^2$ and half of space is liquid $k^{-1} = T^2 \div a^3$ where liquid is moving. However, it is also true that everything through movement defines a value in relation to one point holding singularity k^0 and that is what the formula $k^0 = a^3 \div (T^2 k)$ underwrites. What this proves is that gravity is the motion of space provided by time being the liquid. Please allow me to explain. In the formula $a^3 = T^2 k$ the space forms as the space is in motion. Newton suggested that $\frac{dJ}{dt} = 0$ where he stopped time to have the motion of the circle demolish the work that the circle does. That means he got time standing still or being T^1 and the motion $T = 0$. Let us ponder on that thought for a while, while we remain with the formula Kepler suggested and then it will seem that according to Newton $a^3 = T^2 k$ and in that T^2 then becomes 0. Should that be the case then we have space going flat because $a^3 = T^2 k$ where $a^3 = T^2 \times k = 0$ forming a square instead of a cube, and the Universe we have is a three dimensional system in every aspect there is.

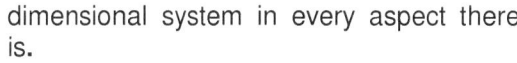

I am of a very different opinion about Newton's point of view where he declared that forming a circle moving $\frac{dJ}{dt} = 0$, and by doing such movement removes Kepler's relevancy factor. This places a value of empty space in which a top would spin and Newton missed the difference there is between a top spinning and a top laying on its side on the Earth. There can be no such a thing as empty space. The fact that space is valid removes an empty connection because space can be anything there is in space. The Universe is time contained in space, which makes it space-time. Space has only one value, and this is to contain time and time provides space with a definite value.

I do not disagree for one instant with Newton's calculations and therefore I am not going into repeating the entire calculating process. All of the calculations Newton made are very correct except the eventual and final conclusion Newton came to. Newton never understood the mathematical concept of time playing a part in physics. In the time of Newton singularity and the relevance thereof had no feasibility in any concept regarding physics. Newton had the concept that time could stand still and that is impossible in physics or any other place. Time can never stand still because time is forever moving by establishing space in a three dimensional environment.

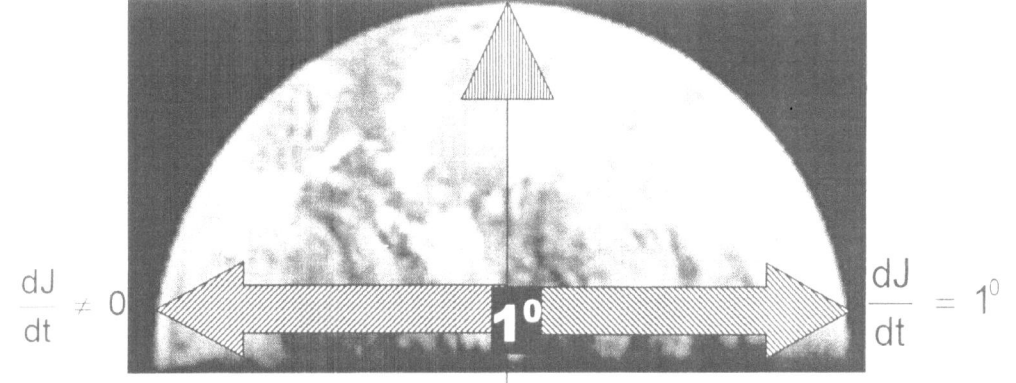

Being the mathematical genius as Newton is so often portrayed as, Newton had very little insight into mathematical possibilities, because when he suggested that $\frac{dJ}{dt} = 0$ he made one huge mathematical blunder. Newton or no other person may place any two objects in a direct relation where the two factors divide and have an outcome that is forming zero. Much surprising is that not one mathematical genius that came after Newton drew the correct conclusion that forming $\frac{dJ}{dt} = 0$ is mathematically not acceptable. Newton saw that dividing something into something else could bring about zero and that is impossible. In concluding that $\frac{dJ}{dt} = 0$ bringing in zero as a legitimate value Newton found a way to replace Kepler's symbolic relevancy value of **k** with using the symbols G $(m + m_p)$. In doing that Newton painted a picture that has no real meaning except where Newton tried and succeeded to put mass into an argument that has no true validity in cosmic principles. This is just a longer and probably a more detailed manner of indicating **k** and better defining of **k** but it symbolises precisely to the point what **k** stands for nonetheless. I wish to draw your attention to the matter of Johannes Kepler's findings that Mainstream science considers as resolved and closed for many a century while it is not. My investigating Kepler helped me to resolve other unresolved matters but it was only possible by using Kepler's work.

I too am well aware that at first glance you will immediately arrive at the opinion that the theme of the book has to be considerably below the standard of an intellectual Master such as the readers must have, due to the position such readers hold, and because of that, the normal research work the readers do. I realise it is dealing with a subject school children learn but in that comes the issue that goes unnoticed. Nevertheless, I hope that this writing may spark interest even at such a low academic level and grade in scientific sophistication and development because I am about to prove that I discovered:

Newton did not think the situation through when he contemplated about gravity. Newton should have thought about factors keeping the gyroscope upright while the gyroscope is spinning. The gyroscope will fall on its side when not spinning and in that position the "Earth's mass" could play a part since the gyroscope fell on its side. However, as soon as the gyroscope started spinning, the balance shifted in favour of a position wherein the gyroscope stands upright. What then comes about has the ability in keeping the gyroscope upright. This is rotational movement and in my other books on the **_Absolute Relevancy of Singularity_** I explain how rotational by the square of the double seven forms Π and Π is forming the curvature of space-time and in that bending of space-time is what we call the atmosphere that keeps the gyroscope square with the Earth and through that the gyroscope stays upright. The gyroscope is acting in accordance with the Coanda effect where the Coanda effect is gravity. By spinning it establishes a solid forming as $k = a^3 \div T^2$ and a liquid forming as $k^{-1} = T^2 \div a^3$. By spinning $T^2 = a^3 \div k$. That is evoking singularity $k^0 = a^3 \div T^2 k$ establishing gravity $a^3 = T^2 k$ in relation to the Earth evoking gravity through also spinning.

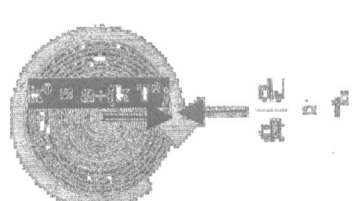

Newton found mathematically that the movement of the top by spin removed the value of the radius $\frac{dJ}{dt} = 0$ where quite the opposite applies. The spin of the top $T^2 = a^3 \div k$ positions the relevancy that **k** as a factor produces by initiating singularity k^0 on both sides of the relevancy $k^0 = a^3 \div T^2 k$ as well as placing singularity in relation to the spinning top $\frac{dJ}{dt} = 1^0$ because that is the correct mathematical principle coming from the equation. Thee smallest any dividing can be is one and one is the producing of singularity. The spin of the circle does not eliminate the relevance of **k** but institutionalise the measure of **k** by confirming the space a^3 in terms of singularity k^0. However **k** has no confirmed and specifically applying value but puts a relevancy of space a^3 forming in relation **k** to movement T^2 applying. By trying to find a measured value applying to **k** is showing no understanding about what **k** is. The value of **k** is finding the space that **k** indicates in terms of what moves. The indicator **k** identifies the space a^3 that the circle claims in terms of singularity k^0 that the movement T^2 isolates from the rest of singularity $\frac{dJ}{dt} = 1^0$. The value of **k** is dictated by T^2 as the movement isolate the space a^3. The measure of **k** is the relevance **k** is claiming on behalf of the space a^3, which uses the relevance of **k** to put a limit the space a^3 spinning in accordance with T^2.

Let us have a look at the bicycle. It is said that the bicycle works on a balance and by science mentioning that the rider of the bicycle is applying a balance, in that they think that the entire problem is solved. That is so typical of Newtonian simplicity about a very complex issue. It is the same as putting gravity down to mass pulling by some small particle called the graviton without ever showing any ability to look more intensely to find a solution for a very complex problem. They always go about by creating a graviton or creating dark matter to look for solutions that solves all the unsolved issues and never do true investigative research in what applies to complex issues such as gravity. Saying the riding of a bicycle is due to balance is the same as putting everything in the Universe down to mass taking charge of particles. The simplicity in which a complex issue is solved becomes as laughable as filling outer space with nothing until the nothing runs over and the Universe that can't gain more because it holds all starts to expand. As the wheels spin (T^2) the relevance of the down thrust **k** leaves the bicycle firmly attached to the ground and in doing that it confirms the space in location (a^3) in terms of singularity

k^0. Newtonians would call this having mass or whatever. Then when having the bicycle moving forward in terms of individual cycling such forward thrust gives a relevance of (**k**) to the peddling power and the movement (T^2) then is about having momentum in relation to the Earth spinning. That means **k** can push down and **k** can push the bicycle forward.

What Newton suggested while never realising he did suggest is the following, and that is that the rotary movement of objects puts singularity $\dfrac{dJ}{dt} = 1^0$ in position on

the outside of the moving circle. However, by using $\dfrac{dJ}{dt} = 1^0$

Newton placed emphasis on the turning movement of the circle and saw this as a destroying of the circle while in fact the turning is putting the space that identifies the circle on the cosmic map. That Kepler also found without ever realising what he found. Kepler said $a^3 = T^2k$ which is $k^0 = a^3 \div T^2k$ which is the spin or $T^2 = a^3 \div k$ which is the circular movement T^2 that validates the space a^3 in relation **k** to a centre k^0 which is exactly and precisely what Newton said when Newton said $\dfrac{dJ}{dt} = 0$ that actually should read $\dfrac{dJ}{dt} = 1^0$. The

location where Newton placed singularity as being singularity established by the movement of space $\dfrac{dJ}{dt} = 1^0$ I

named **eternity** because there nothing can ever go bigger or become more. Whatever was and is and will ever be is locked in that space I named **eternity**. The "so called expanding" of the Universe $T^2 = a^3 \div k$ is where singularity is shifting relevance **k** from liquid $k^{-1} = T^2 \div a^3$ to solid formulated as $k = a^3 \div T^2$ and the process whereby this happens is precisely the same as the Coanda effect. Getting back to my first argument about a line and that no line can start at zero but has to use singularity as a starting point, this is all the proof I require. The line **k** coming from the centre (singularity k^0) forms by forming an initial dot Π^0. However, I went on to say that whatever the line used to start with has to continue in order to repeat the same that began the line. Therefore the line started with Π^0 and it has to continue with Π^0 until such a point, as it must end with Π. Whether the line is Π^0 or is r^0, or uses 1^0 the outcome all refers to singularity being used. By reducing the line we come to the end of the mathematical equation of the circle but the circle does not end there. That is what Newton did not recognise from the figures the cosmos represented to Kepler. The circle only secures the final cosmic figure and the value to singularity where all things have equal value. The movement of the circle splits singularity in two sectors. By forming Π the circle has to form Π^2 due to the movement coming about in securing the space Π^3.

Kepler chose to use different symbols too those being valid, but the concept remains the same. Kepler said that $a^3 = T^2k$ while I show that $\Pi^3 = \Pi^2\Pi$. It still confirms that movement $\Pi^2 =$ is the forming space by three dimensions Π^3 in relation with the movement Π^2 being relevant Π to singularity Π^0.

I shall try and explain what this concept holds in terms of a piston moving while working inside an internal combustion engine. The piston goes up to a point we call top dead centre where the piston stops and according to the crank the piston halts in directional movement. Then the piston starts to accelerate to a point we call bottom dead centre where, again it comes to a dead halt. The piston stops directional movement at T.D.C. and at B.D.C. or that is what we see without seeing anything. This is not the case because if this was the case the engine must vibrate at those two points of stopping. We reason that the piston stops twice and starts moving on the two occasions but if that was the case of stopping at two points without stopping anywhere else, the vibration that the stopping will cause will have the engine disintegrating completely. To us favouring positions the piston stops at two locations but the fact of the matter is that the crankshaft stops every 7° of rotation and if the crankshaft stops, then so does the piston stop.

The stopping is a continuous and is an ongoing process that happen every 7° of rotating. The crankshaft moves in a straight-ahead position going straight and then it stops and redirects by 7° and then it turn by going straight again. It is $a^3 = T^2k$ and then it stops (a^3), it turns (T^2) and then again goes straight again (**k**) while holding reference with singularity $k^0 = a^3 \div T^2k$ all the time. One cannot part the redirecting and the going straight T^2k because it is the same movement since the space forming a^3 is equal = to the turning T^2 and the going straight **k**. This is evident when dissecting Kepler's formula $a^3 = T^2k$ that $T^2 = a^3 \div k$ and $k = a^3 \div T^2$ while honouring Newton's 3rd law $k^{-1} = T^2 \div a^3$. Please believe me that this puts movement is a much complicated dimension because this has the material $a^3 = T^2k$ moving $T^2 = a^3 \div k$ in terms of ($k = a^3 \div T^2$ as well as forming $k^{-1} = T^2 \div a^3$) while always referring to singularity $k^0 = a^3 \div T^2k$.

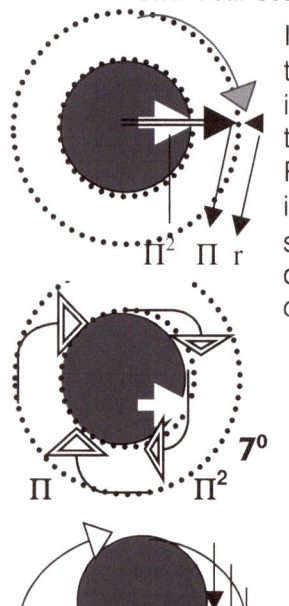

Π^2 Π r

Π 7^0 Π^2

In the circle $\Pi^2\Pi$ the space surrounding the rotating object will also extend by Π as the concentration of the spinning motion draw or drag on past Π^2 extending the influence of Π^2 by the value of Π. This extending of Π^2 to accommodate Π we refer to as the atmosphere, but physics apply to this extending in the normal fashion. From the spinning motion Π does not stop at the end of the solid structure but the influence of Π extend and this then becomes the atmosphere. The influence of Π^2 stops at the end of the solid structure but the influence of Π extending plays a most dominant role in the cosmos, although not yet recognised and that factor is most crucial to a better understanding of the implications of laws governing the cosmos.

On Earth we can measure a distance between two objects to a precise measurement and come back the next day to find the two structures well secured in the same place and distance apart. That how ever we also know will not be the case in space since the objects will always drift apart and away from the position it had. With the circle being $\Pi^2\Pi$ the Π^2 will reflect the circle in the square with Π forming the extending of Π^2. The extending of Π will not end immediately but will carry to the surrounding space the circle influence through rotation. It is clearer than ever before that the principles of " gravity" distinctly differ from the Earth surface to outer space.

Every quarter is directly opposing the next as well as the previous quarters thereby starting a new set of principles it has to adhere too, but breaks by moving through time anyhow.

In that movement comes about destruction of the self-preserving because any change of what ever small proportions will lead to destruction coming about as if with a snow ball effect. The top can be its personal matter and anti matter just by changing the speed of rotation where the points does not precisely meet the previous points and deformation stars by overheating bringing about an altogether change in relevancies to itself as well as other matter in the same orbiting time.

With all matter having the same start from the same singularity, all matter should therefore be synchronised in growth and in rotation, where the matter is in support of all surrounding matter spinning around the common and original singularity that produced similar growth and rotation speed since time began to the present day.

The Roche limit in the practical sense.

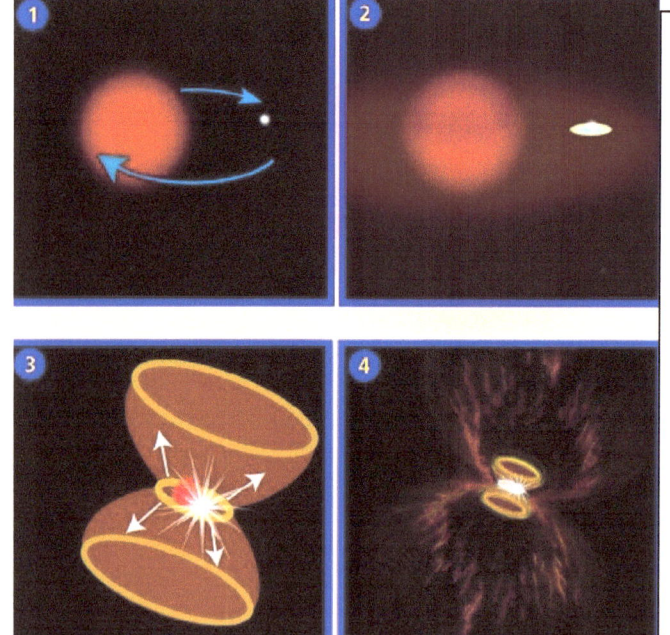

In the first picture the major partner stars spinning the minor partner ever faster. This is because the line $\Pi^0\Pi$ extends to connect to the minor stars value of Π and by connecting it increase the movement of the minor partner's value of Π^2. Taking control of the minor star's gravity Π^2 it takes hold of both the factors 3 as well as 4. In singularity the line as well as the circle not only images the major star but by taking control the major star takes lover the value of the system of then minor star. It HAS to be not the same but it is the very same since the dot in the major star is the dot in the minor star (1=1) or ($\Pi^0=\Pi^0$) and with that it expands the space $\Pi^0\Pi$ of the minor partner and thereby liquefying the density of the minor partner. The liquid layers get expanded to become gaseous and become the atmosphere material of the major star. This then leaves the inner solid layers of the minor star exposed and not feeding off the gas layers. Then that lets the inner solid core overheat.

By not feeding cosmic gas turned to cosmic liquid the inner core of the minor star overheats and expands far beyond the endurance the core is able to withstand. With that the Roche limit comes into effect by dividing the core into singularity sectors of $\dfrac{\Pi^2}{4} = \dfrac{(\Pi)}{2} \times \dfrac{(\Pi)}{2} = 2.4674$ and that we can see in the mann er in which the core divides into the two sectors depicted in the picture.

This is not only limited to planets in our solar system. In the Universe, there are giant stars spinning around each other. These stars are binaries, which are also one form of double stars where double stars are another such a form. The difference between the types depends on the distance they remain apart. They keep a certain distance apart and do not collide. In the case of the Sun and its planets, it could be a case that the systems might be too small, or they might be too apart. However, this is not the case with binary stars. They are close, they are big, and they spin around a mean axes called the Roche limit.

Since occupation may or may not be placing the factor in infinite, the space therefore holds the premier singularity of infinite from which all included in the Universe has come.

When the top starts spinning in a specific position the top merely executed the option to fill the premier singularity at that specific point. When it moves it may take the premier singularity with to the new location it moves through spin or it may fill yet another position in singularity as all is the same.

Because every atom spins, singularity being in every atom, singularity is in every atom surrounding the singularity. With every atom becomes the centre of the Universe holding all outside singularity as space-time and by conclusion of the cosmic splendour, all space-time will dissolve once more back to singularity, bringing about the last atom holding all singularity without any surrounding of space-time left.

The influence immediately above the circle will have the biggest influence and reduce gradually as the value of Π reduces in the leverage that the space has on Π and a gradual but definite change from Π to r will affect the extending of Π progressively more. The decline of Π will follow the same contour of the circle at 7^0.

From there it influences singularity in the triangle flowing through to the half circle. It is an interaction between circular and linear motion as the value of Π continuous past Π^2 (at the end of the solid) and every cosmic structure holds an individual and specific singularity. The field where Π extends we call the atmosphere having a value of 21.991 / 7, which is Π.

From this line of reasoning I dismissed the theory of the presence of a force being gravity but rather consider it as a dimensional changing contributed by the spin of the Earth and the spin comes from singularity located in the centre of the Earth.

By claiming the position held by singularity premier as a vacant spot until the arrival of the top, the singularity of the top divides the point flowing from singularity into four sectors holding two half circles

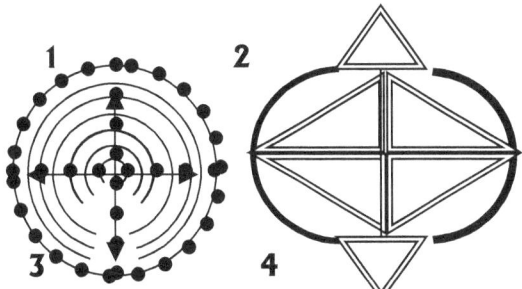

From the star holding a dominant point or most valued point in singularity it affirm all three other structures, each holding singularity individually and in a compliment of 5.

The Roche limit explained in the Newtonian view and showing they have no idea how it functions and by which they clearly don't make much sense about how or why it is there at all.

The Roche limit is:

The region surrounding each star in a binary system, within which any material is gravitationally bound to that particular star. The boundary of the Roche lobes is an equipotential surface, and the lobes touch at the inner Lagrangian point, L_1, through which mass transfer may occur if one of the components expands to fill its lobe. It names after the French mathematician Edouard Albert Roche (1820-83).

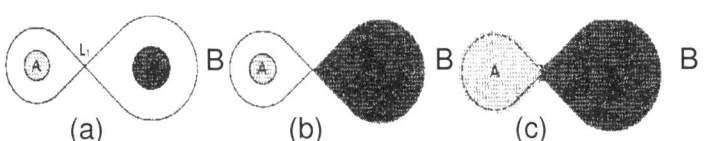

THE ROCHE LOBE: In a binary system, the Roche lobes of components A and B meet at the L_1 Lagrangian point. (a) In a detached system, neither star fills its Roche lobe. (b) In a semidetached system, one massive

component, B, fills its Roche lobe. (c) In a contact binary, both components overfill their Roche lobes and share a common envelope.

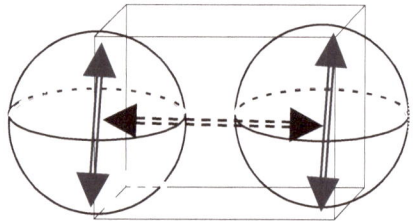

Radius formed by singularity carried

Π

$\Pi \div 2$

$(\Pi \div 2)^2$

$\Pi \div 2$

Π^0

In the Roche limit the space factor provides space to a solid structure and therefore the value of r is replaced by the value of Π bringing about a half of Π square. The cube holding 5 to either side removes an extending value allowing the extending of Π to indicate position to space by movement and that would be a square. Where Π extends to lock onto the next sphere's extending indicator, Π has to connect to Π forming the square of space and translating that to the half of Π being $(\Pi/2)^2$.

It begins where the cosmos began and that was when singularity performing as infinity parted from singularity performing as eternity. I am not spending time by going into detail on this mater because of the enormity the subject holds but I have written an entire book explaining how this process came about as it used the four cosmic pillars to generate the first gravity. But as space-time grew, the extending of the influence of the four pillars grew while still holding onto every detail that applied in the first instant time became space. The line that forms is still presenting singularity as a line Π^0 and this value supporting singularity runs across and through every atom forming such a line.

At the end where everything is the centralised $k^0 = a^3 / (T^2 k)$ the end holding singularity Π that forms by movement $T^2 = a^3 / k$ that validates space $a^3 = k\,T^2$ forms a specific relevancy $k = a^3 / T^2$ that goes both ways $k^{-1} = T^2 / a^3$. At such a point Π forms one half reaching into time in order to expand into time while the other half moves down as space or matter pulls it to compress it. This effectively splits Π into two halves and by Π going into a square; it is half of Π going square which is $(\Pi/2)^2$.

5/2

Five sides divided by two spheres.

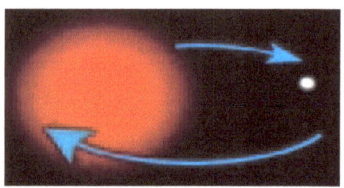

7   Space-to-matter

The Roche limit
5/2 = (Π / 2 X Π / 2) = 2.4674

Explaining the "not understood part" is what is really informing information about the cosmos.

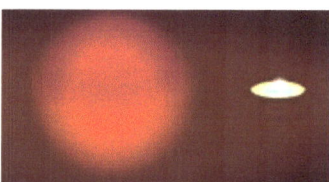

This is how the Roche limit functions in the practice where governing singularity takes control.

The formula $F = G \dfrac{M_1 M_2}{r^2}$ cannot explain the behaviour of stars going into a melting frenzy such as the pictures indicate because $F = G \dfrac{M_1 M_2}{r^2}$ **presumes that stars has to collide and not come onto battle while the stars are still a large distance apart.**

The obvious occurrence shown in the pictures above indicates clearly how the major star liquefies the minor star and takes control over the singularity dynamics within the minor star. By understanding cosmology principles correctly I can explain what is occurring in this instance and this occurrence connects directly to the Roche limit, as explained above. Not only does the Roche limit explain this phenomenon, but also it ties directly to the Titius Bode principle, but also it rubbishes Newton's formula $F = G \dfrac{M_1 M_2}{r^2}$. According to the science formula of

$F \;=\; G\,\dfrac{M_1 M_2}{r^2}$ the orbiting structures should collide with a bang as r^2 diminishes under the pulling power of the combining force that both objects hold in their mass. Instead they do the tango until one drop, but when dropping it still does not collide with the larger structure, as would the formula used by science, $F \;=\; G\,\dfrac{M_1 M_2}{r^2}$ suggest. The major star liquefies the minor star and when the minor star is in a complete state if liquid, the major star absorbs the minor star by applying gravity as the cosmos uses gravity through the Coanda effect.

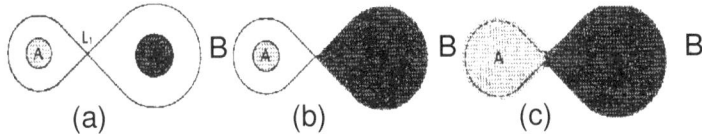

(a)   (b)   (c)

THE ROCHE LOBE: In a binary system, the Roche lobes of components A and B meet at the L_1 Lagrangian point. (a) In a detached system, neither star fills its Roche lobe. (b) In a semidetached system, one massive component, B, fills its Roche lobe. (c) In a contact binary, both components overfill their Roche lobes and share a common envelope.

Once more, this phenomenon should not occur with Newton's presumptions about gravity. These bodies will collide and destruct, without a doubt. When the formula $F \;=\; G\,\dfrac{M_1 M_2}{r^2}$ applies, there should not be any force, which is able to keep them apart. However, they do exist and what is more, they maintain a certain distance apart. Seen from this view, it is little wonder that the significance of this was lost in the notion that this is yet another "mystery" of the Universe. The scientists of the day (and the past) lost the importance, which this holds for us as Earthly dwellers. Newton used the top to point out why he thought mass was the uttermost important factor that provides gravity. I wish to use the top to point out what Newton missed.

Looking at the top when spinning it seems that the body structure of the top is solid and the air surrounding the top is liquid. It also seems as if it is the air that is supporting the erect stance of the top. The top as a structure composes of solid particles that light cannot penetrate and that material cannot pass through. In that sense it seems to fit all the conditions we set for solidness. The top spins by contracting the air in the direction of the top and it spins through the compressed air that allows the top to spin while being erect seeing that the top air immediately surrounding the top has much more density than the rest of the atmospheric air has.

That which can move is the liquid part of singularity, which forms singularity by the air surrounding the top. This then is that part that moves and that stands related to what cannot move being singularity within the centre of the top. Since everything is singularity everything is immovable but also since everything is singularity that which does not form part of that which can't move forms par of that which can move. It is Π bringing the partition by instituting Π^2 in relation to Π^0.

Every thing outside that which forms the body of the top is liquid with the top forming a solid or so it seems to us. Well yes in a way and not that much either. The top through the atoms that constructs the body of the top is a pump that pumps heat from the outside inwards just like a turbine engine. Every atom that is rotating inside the structure of the top is keeping the centre erect.

The centre is totally motionless because all the atoms in the top are moving with the spinning body while in relation to the centre all the atoms are holding a position being uncompromising still as solid the solid body's atoms never changes their local position and therefore they never move in accordance with the centre. With all the atom centres remaining rigid and still in one position as witnessed by the centre singularity, therefore the moving of the top circle goes to the part forming the liquid outside the body of the top since seen from the centre that is the space that changes rapidly with motion. The movement is thus extending the singularity in movement of the

Singularity continuing to the value of Π^0

Π^0 •••••••••••••••• Π **Singularity also continuing to the value of Π^0**

It is the movement of Π^2 that alters the line of singularity parting material from time Π^0

top to the outside of the edge of Π forming Π^2 where the top meets eternity. The extending of singularity is holding the air as a liquid and being the liquid flow of the liquid keeps the top erect and spinning. The spin produces a cold in relation to the hot that the liquid is. From this we can see that although it is the top that is moving, as seen from our perspective and seen from our reality the opposite applies as seen from the centre of the top the top. According to singularity coming from the centre of the top it is not spinning because the motion is transferred to the outside of the top. What is moving is liquid and what is not moving is solid. Everything has a reference in relation to another point. That which is capable of relocating is forming a liquid in relation to that which is securing the position of rotation. Everything in the cosmos can move and yet that is only true when there is one point in the cosmos that at that moment in reference are not able to move. The cosmos stands divided between the eternal moving of eternity and the immovability of infinity.

Everything around the top is liquid with the centre being a solid. However the solidness and liquid has cosmic standards and just as it is in the case of hot and cold, big and small, fast and slow, our standards and cosmic standards do not share any measurements. So too does cosmic notions about liquid and solids have a totally different meaning in cosmic terms.

There is a pumping interaction of space-time flowing towards singularity through every point that confirms singularity. Everything in the top that forms the material is also surrounded and partitioned by liquid. By providing motion the matter in the top serves forming the solid connection with the centre as the liquid factor moves towards the centre and in that it extends the space that singularity provides. The structure is composed of atoms. In the atom there is a governing generated singularity around which all sub-atomic material rotate. In the case of the atom all the rotating material forms the heat while the generated centre, which is incapable of rotating, forms the solid factor. Every aspect that is without motion stands in a relation of 1^0 and that which is relatively moving or changing location or finds a new position holds 1^1. These are names I have given to bring some understanding to everything being equal anyway. Everything that is standing still is 1^0 and everything that is moving is 1^1.

Gravity or motion is a constant relation that solids have with heat where heat forms the liquid and solids form space. There is the rotation but part of the rotation is the lateral progressing by rotation to also confirm the generated centre by relocating the entire centre in a straight line. The generation of the line forming singularity is vested in the rotation but the flow towards is the lateral forms another factor motion brings about just as electricity produce a flow of time in relation to space collapsing.

The forming of space-time by measure of gravity is using the same system to do the very same everywhere. Kepler said it to be $\mathbf{a^3 = T^2k}$ while Newtonian science portrays it as $F = mv^2$ which when formulated correctly should read $F^3 = mv^2$ and Einstein formulated his version as $E = mC^2$ which also if formulated correctly should read $E^3 = MC^2$. Is there anybody that is brave enough to show how that which followed Kepler was formalised in any other way than Kepler's product given the condition that the mathematical interpretation should read correctly.

There is no substance difference between 1^0 and 1^1 and it is a relation where one moves as the liquid partner and the other stands still as the solid factor. Both are not as much equal as they are precisely the same. Infinity cannot move and eternity cannot stop moving. By parting, infinity had to remain motionless and eternity had to remain moving as it introduced a part of the cycle to one line (point) where it stops moving in relation to the other factor that cannot move but does start moving. Time is a graph that never begins since it never ends and while everything never repeat everything always never remains the same. The factor that shows motion forms the liquid while at that moment the factor that does not show motion forms the solid. The measure of 1^0 is transformed to 1^0 and which ever are 1^1 is passing the extending of space on to 1^0.

Time spins around a centre because everything spins around some centre somewhere in order to secure the centre singularity. But also time moves and in that there is the linear that always are part of cosmic motion. The centre is referred to by heat but heat also secures the centre by reconfirming the centre in the lateral. But in both cases singularity is reinstating singularity by confirming it as it is referring one another. In the manner that 1^1 confirm a position in singularity 1^0 it is supporting 1^1 by generating 1^0. By generating 1^0 it is repositioning and reallocating 1^1 as a position by confirming 1^1.

From all of the above one can deduct that outer space is something viable through which objects travel. It is clear that something filling the space between Jupiter and its first moon because of lightning interaction between the two structures. There is a reference between a structure holding material and the space above as well as the space between two objects. There is electric lightning travelling between Jupiter and its closest planet. If there is lightning there is electricity and electricity means a very distinct interaction that connect the space inside the material structures through the space parting the material structures putting the space in between as a conducting medium which nothing can never do. Understanding this nothing not existing is fundamental in understanding the relevancies applying.

Considering the notion of nothing being in place filling the space between the cosmic structure, electricity needs a conductor to transmit the interaction there are and that disproves the nothing theory and it puts the nothing science places in the Universe as nothing those scientists have between their ears in their understanding the cosmos. It is official that the interaction was detailed as $a^3 = T^2 k$ which is what Kepler found, yet with this information science still do not appreciate the fullest of the implication. I have changed the formula to a^3 (space) = T^2 or gravity (time) k (the relevancy applying) and this becomes singularity $\Pi^3 = \Pi^2\Pi$. By applying the atomic value the relevancy changes to $(\Pi^2 + \Pi^2)\Pi^2\Pi3$ and that relevancy projects to cosmic atoms such as two stars interacting. When two objects come closer than the relevancy would permit, cosmic laws change their application and in this case then becomes $(\Pi^2 + \Pi^2)$ from either side where the three of space changes to singularity Π acting as the influence $(\Pi/2)$ from both sides making that influence a square $(\Pi/2)^2$. All of this I explain in the Cosmic Code and I am not dwelling into this matter at this stage.

There is a battle for position of dominance between closely allocated stars where the rivalry that Binary stars ad in that case "mass" pulling is proven none existing. The Binary systems never collide as Newton's formula $F = G \dfrac{M_1 M_2}{r^2}$ insists on applying through out the cosmos. The closest destruction takes place is when a star is 2.4674 times the diameter of the major star. Any person with the last understanding of mathematics would recognise the value of 2.4674 as $(\Pi / 2)^2$. This indicates to the presence of singularity governing the area $\Pi^2/4$ where both object will hold onto their individual singularity while spinning around the mutual point of singularity.

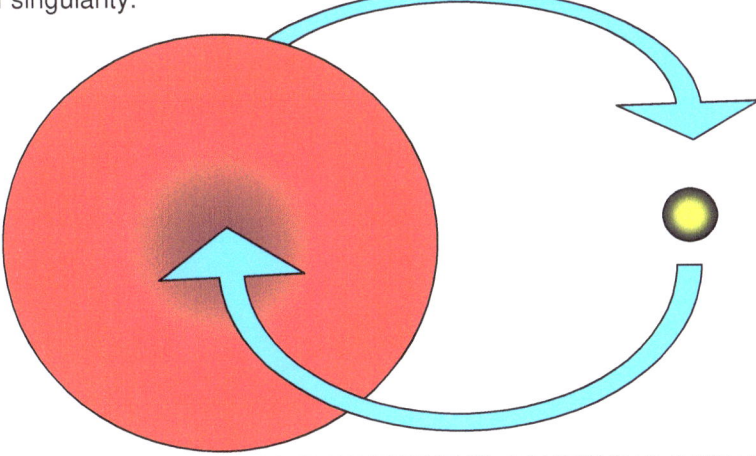

THE ROCHE LIMIT FORMS AT A LINE WHERE THE GOVERNING SINGULARITY OF THE MAJOR STAR CLOSES THE BORDER THAT MAINTAINS THE APPLYING SINGULARITY INTEGRITY.

Where there is two stars in close proximity the one star is unable to defend the singularity governing the structure as to regulate the relevancy of self-conservation, then the major star will take control of the minor stars governing singularity and destroy the singularity by establishing overheating with in the singularity of the minor star. I think NASA refers to pictures depicting this as "blowing bubbles or blowing heat " but do not quote me on that. I just found it amusing at the time that the beast brains in the world would come up with nonsense like that.

To the one object, anything distant from its proton cluster is space, space in what- ever form. by destroying the singularity, the space becomes heat either under its influence or under its control. On the one side, space holds the value of Π and on the other side, it also holds a border of Π. The time however changes to defend the singularity therefore times the square of the atmosphere applying then holds position where space normally holds position.

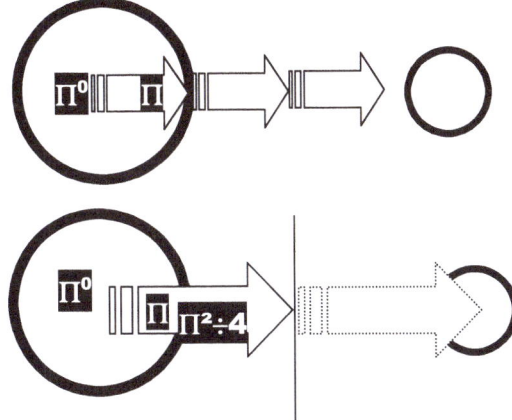

Anything spinning alternates its position in relation to the circle four times. This produces the invert square law. In a developing binary there is one major and one minor component. The radius that forms holds a cosmic relevancy of Π^0 forming Πr^0 that then by movement forms Π^2. The full circle of rotation will be 4 (the 4 directional changing quarters) that are related to Π^2. If there is a structure within the $4\Pi^2$ it is going to interfere with the circle that the major star ahs to maintain in order to rotate. In order to maintain self-preservation in its rotation the major star will clear the circle of $4\Pi^2$.

In order to se to the fact that the rotation circle is clear, the major star will take charge of the governing singularity of the minor star and alter the relevancy of the governing singularity ($\Pi^0\Pi$) to fit the relevancy applying of the governing singularity ($\Pi^0\Pi$) within the major star. This will in turn alter the gravity ($\Pi\Pi^2$) of motion of all the atoms within the minor star. By liquefying the lesser star had to point to the interaction there is in the cosmos between solids and liquids. This gave rise to my theory of relevancies swapping and fluids being consumed by solids. What happened during this interaction lead me to believe that the Coanda effect forms gravity and this has nothing to do with having mass or not having mass.

LINE HAVE DUAL OR MUTUAL SINGULARITY

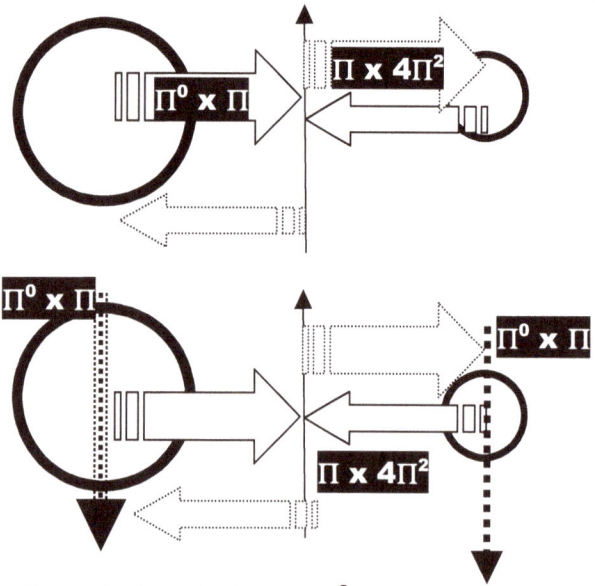

The mutual line of singularity $(\Pi/2)^2$ holds a position referring to each objects individual line of singularity in the value of $(\Pi^2 \div 4)$

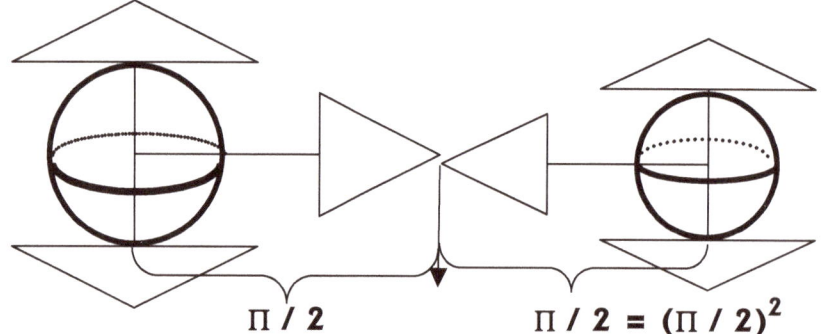

$\Pi\,/\,2$ $\Pi\,/\,2 = (\Pi\,/\,2)^2$

SINGULARITY MEETS AND COMPLIMENTS EACH OTHER.

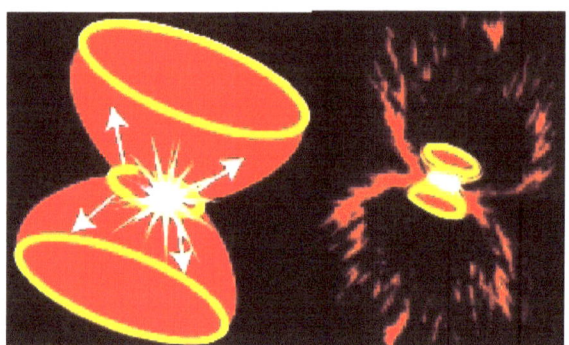

In the struggle for superiority begins between the two points each holding a spot of governing singularity. This is the very same as what takes palce between the spinning top and the Earth or if you wish the Earth and the flying supersonic aircraft where the Earth is the obvious eventual winner and in the final outcome both objects maintain fighting for the claim to singularity, by pushing the space-time occupied to new levels of occupied space-time values.

The result of this is the establishing of an individual singularity dominating because of the unequal ness between the points of singularity and by cosmic law liquid is always present in relation to solid. The minor star is liquefied to bring the cosmic law into effect where at a rotation distance of $(\Pi)^2$ first favouring the on in the matter part and the other in its space part and afterwards turning the points of reference around.

Material moving brings the divide between eternity ands infinity by referring to singularity $k^0 = a^3 \div T^2k$ as Kepler's formula $a^3 = T^2k$ being $k = a^3 \div T^2$ while honouring Newton's 3rd law $k^{-1} = T^2 \div a^3$ forming gravity or time as singularity spinning $T^2 = a^3 \div k$. **Gravity and movement and time is all the same thing. It is repositioning everything in relation to one point holding singularity. That brings us to movement.**

It is Π that sets the margin as to where the material starts and time or outer space begins. It is Π that sets the limit between that which is liquid and that which is solid. It is the positioning of Π by the movement of Π^2 that part infinity from eternity. It is Π that ends that which can never begin and begin that which can never end. It is Π that establishes a reference point where singularity divert into the eternal and the infinite allowing matter the zone matter can claim space in time, the control of space in time and the influence on space-time, the point of singularity have to reduce the value of singularity on both accounts of the cosmic atoms claiming individual singularity, or enlarge the claim of space by matter away from singularity. This is rather important to understand when arriving at the actual presentation of the formation of the solar system

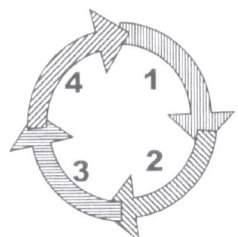

Any point will be opposing itself within the **rotating of 180°** where it **then change every aspect** of its **previous flowing** characteristics it had or **will once again have** in 360⁰ from there. While in rotation from the view point of a bystander it all may seem static and never changing but to the object in spin every next instant in time will be diverting from every aspect it had every second passing, and the direction it held in relation to the direction it held the previous mille, mille second will totally be incompatible with the direction it holds the very next mille, mille second of rotation. In this fact of an ever-cyclic change going on forever hides all the mysteries about "global warming" and all the misinterpretations persons such as Al Gore and others attach to "global warming". In ever directional changing of 5 to 5 we have the 10 of gravity coming from 2 x 5 as 7 spins through 5 and the change in direction by seven confirms five twice.

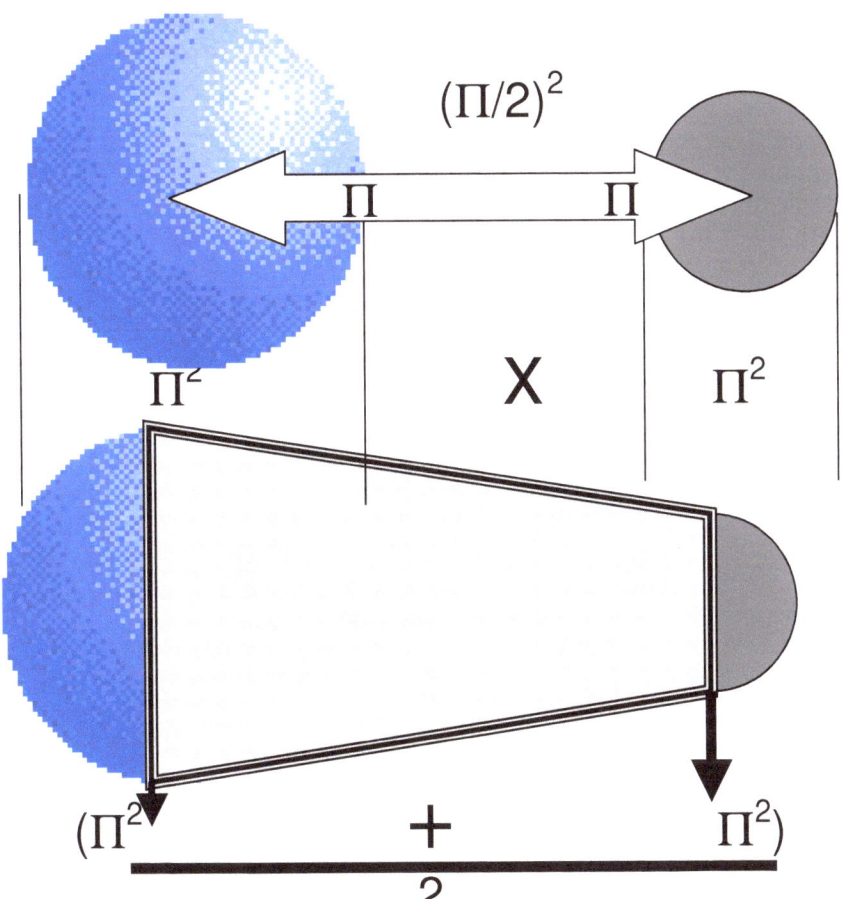

The sectors provide individual singularity a means in sustaining governing singularity by which provision comes through maintaining governing singularity the required spin in maintaining cooling. If this process did not apply, there would be no connecting individual singularity to major singularity.

Because there is a space that may not be occupied by a particle does not exclude the possibility of a particle sometime to the future occupying that space. If the space was nothing all possibility of future occupation will become excluded by the presence of zero that is unable ever to include occupation. I have to eb eprsistant on this fact that zero does not form outerspace because my work was rjected on several ocations because sciewnce insist on the madness that geodesic outer space holds a firm value of zero.

From the centre of the top runs the premier singularity and as the top starts rotating the top's rotation bring about the sides to singularity, which too was present all the time but filled the being, they're by the rotation of the top. Such rotrion involves four specific quarters.

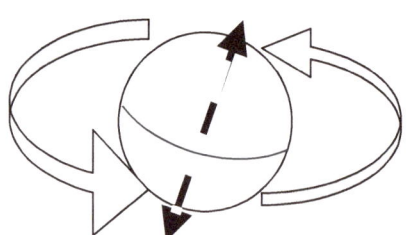

The drawing is the circular Π^2
The movement is the linear r
The change over of dimensions is Π
$\Longrightarrow$ r^0 meets Π $\Longrightarrow$ Π^2

In the action of the inseparable drawing closer and moving closer gravity finds the dual value of linear and circular gravity. There is no separation of the two factors acting as one but both have different application and values in the unit. This is the result of singularity having three parts acting as one but giving three distinctions in application.

In both stars $(\Pi^2+\Pi^2)$ $(\Pi^2\Pi)3 = 1836$ would apply as a relevancy but the two will apply each according to the gravity within. Within the boundaries of the major star the gravitational relevancy will be A = $(\Pi^2+\Pi^2)$ $(\Pi^2\Pi)3 = 1836$ and within B the same will apply in accordance to that gravitational relevancy applying giving the atomic singularity as B = $(\Pi^2+\Pi^2)$ $(\Pi^2\Pi)3 = 1836$. However since gravitational relevancy is not equal in both stars we have the fact that {A = $(\Pi^2+\Pi^2)$ $(\Pi^2\Pi)3 = 1836$} ≠ [B = $(\Pi^2+\Pi^2)$ $(\Pi^2\Pi)3 = 1836$]

Gravity is the dimensional changing of heat holding r as reference to the sphere holding Π as the reference. Heat occupying space has the cube that can apply r, as a straight line bringing about the cube with all its other names than may find attachment to specific form but nevertheless still remains only a six-sided cube with angle changing in some cases. When atoms are within the structure of the Sun, all the atoms form a combining legion that we humans are unable to measure or to witness. In the centre of every atom forms a governing centre singularity. The worth of this are vested within the centre governing singularity of the Earth. This gives the atom's forming the Earth a cosmic code of $(\Pi^2+\Pi^2)$ $(\Pi^2\Pi)3 = 1836$. I am not explaining this code in this book but I do explain it in the Cosmic Code.

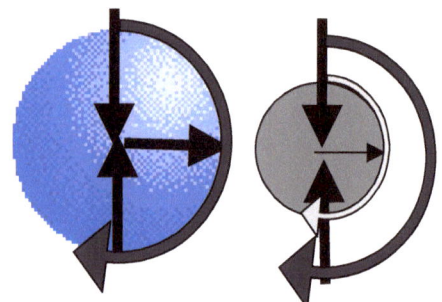

In every star the relevancy of the proton to electron ratio applies differently although the electron to proton ratio seems to equate the same and is the same $(\Pi^2+\Pi^2)$ $(\Pi^2\Pi)3 = 1836$. That is the atomic displacement difference there is that space-time is reduced from the electron to the proton. I explain this ratio in much more detail in the Comic Code.

Being in the star or planet the ratio will have a different defining ratio because of the gravity of every star holds a different value which is determined by the governing singularity vested in the centre which is derived from the combination of the total of all the singularity centred within every atom forming the line Π^0 to Πr^0. It is when this ratio can't maintain an atomic equilibrium throughout all the layers in the star that we find a Super Nova blowouts occur in one or more of the layers of such a self destructing star.

It is the presence of the relevancy factor **k** that is responsible for the atomic occupying volumetric size a^3 as well as the gravity T^2 within every star within every atom in each independent star. The differentiation is the result of the governing singularity, which is the result of the combined value of all of the atoms forming the line that makes up Πr^0. That is what Kepler's formula o $a^3 = T^2k$. The curve of Π runs as a divide all along the outer rim of every star and this divide of Π is derived from every value that every atom forms by placing an atomic Π into the equation.

Again I repeat what I so often repeat when I refer to gravity in my own incapable and uneducated way: Whatever gravity is, gravity has to beΠ. If gravity is linked to mass as Newton stated, then mass has to be very closely connected toΠ in order to assume the role of being responsible for gravity. Looking at every aspect that forms gravity, it is formed by a circle that forms a circle in return. Gravity is the instalment of oneΠ initiating the followingΠ. The curve the Earth holds forms aΠ, which is responsible for the circle the Moon follows as the Moon circles by orbit the Earth. The Earth as much as the Sun as much as all stars and galactica holding gravity is round and the roundness areΠ. Taken much further into more philosophy we find the curvature of space-time, the fact that gravity bends light into a curve, this bending comes in the form of a circle that is formed byΠ. The Sun for instance spins around and that is formed byΠ. The Earth holds the Moon captured while the Moon circles around the Earth and the circle is a result ofΠ. If it is with gravity that the Moon circles around the Earth, then in all of this we must locate gravity holding Π as a value. It is this pattern hat I recognised when I formulated the Cosmic Code. I will give one example to indicate the gravitational worth ofΠ. The Universe starts to go three dimensional at a point where Π forms a three dimensional system which forms at a point $7/10$ $(\Pi^6)6 = 112.16$ which is where the element table starts forming atoms of various relevancies running from $3((\Pi^3) = 93$. This I explain in the Cosmic Code.

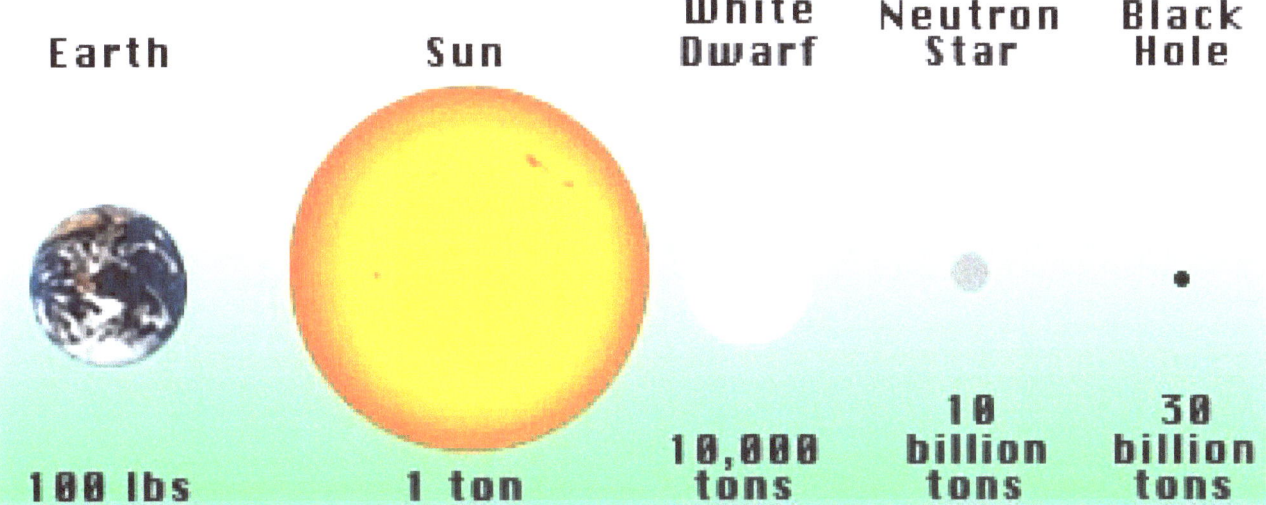

It is clear that the more "massive" the star gets, the denser the star becomes as the star decrease in volumetric space used. The star becomes considerably more compact as its "massiveness" increases, which can only reflect on the situation controlling the atom that controls the stars. The star is its atoms and the atoms control the governing singularity as much as the governing singularity takes charge of the compliment of the atoms in the relevancy value applying as$(\Pi^2+\Pi^2)$ $(\Pi^2\Pi)3 = 1836$. In the more "massive stars" there are more atoms but the increase in atoms forms a reducing of occupied space because the relevancy applying condenses the atom's space-time occupied.

It is the compliment of atoms that form the star and it is the relevancy of the atoms within the star that forms the relevancy of the gravity applying within the specific star meaning that no star ever will have the precise relevancy of any other star. The relevancy applying in the star ($k = a^3 \div T^2$) and (k^{-1} = $T^2 \div a^3$) forms the space –time ($a^3 = T^2k$) that is responsible for the gravity in spin ($T^2 = a^3 \div k$) taking charge and all the while it relates to the governing singularity ($k^0 = a^3 \div T^2k$). It is an interwoven network of interacting relevancies all complying to the governing singularity that is attuned to the controlling singularity. The smaller the circle is formed by the gravity $T^2 = a^3 \div k$, the more compact will the space-time be ($a^3 = T^2k$) and the more control ($k = a^3 \div T^2$) and ($k^{-1} = T^2 \div a^3$) would singularity ($k^0 = a^3 \div T^2k$) have on the star. If this is put correctly in terms of Π applying as it should then it will be as follows: The smaller the circle is formed by the gravity $\Pi^2 = \Pi^3 \div \Pi$ and the more compact will the space-time be ($\Pi^3 = \Pi^2\Pi$) and the more control the relevancy ($\Pi = \Pi^3 \div \Pi^2$) and ($\Pi^{-1} = \Pi^2 \div \Pi$) have by committing the governing singularity ($\Pi = \Pi^3 \div \Pi^2$) and ($\Pi^{-1} = \Pi^2 \div \Pi^3$) have on the star.

The relevancy ($\Pi^2 + \Pi^2$) ($\Pi^2\Pi$)3 = 1836 is valid only as it applies within a star and has a different meaning in relation to what the governing singularity applies. Everyone walks the Earth with the idea that the big stars are those charging "massive" gravity and the small stars has very little gravity because it is "mass" that brings size and the big stars have big sizes and therefore have big gravity fields. That is a load of rubbish and the Hertzsprung-Russell diagram, which is as much rubbish as the idea that mass that produces gravity.

In that there is no mention of mass simply because mass don't feature where gravity is mentioned.

Looking at the Solar system we find that all planets and objects not classified as planets and all things that is just simply forming solar debris has one thing in common…all apply the value of Π in the process where they orbit the Sun, which also uses the formation value of Π to construct the roundness the Sun has. Gravity has much more in common with Π than it will ever have with mass that is producing gravity. Wherever singularity forms gravity, it involves Π which then results in gravity manifesting as some or other form holding Π as a major factor.

However, it is important to note that **k** is not a value but it is a reference. The value of **k** or if you wish then use Π is in determining the gravity factor T^2 or if you wish then use Π^2. It is to indicate amongst many things the atomic relevance applying in a specific space a^3.

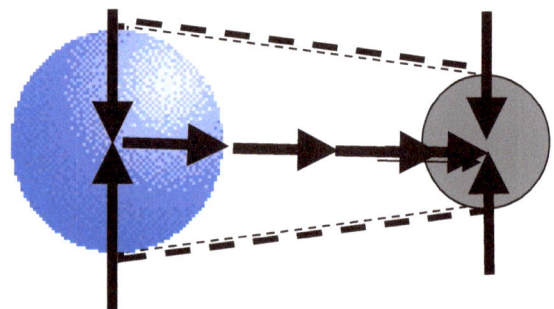

It is clear from the Roche limit applying that the governing singularity influence stretch at least as far as $\Pi^2 \div 4$, which then is 2.4674 times the radius of the star. When observing the Mon we can see the governing singularity influence stretches even further since the Moon uses the Earth's governing singularity as its own axis. We know that $a^3 = T^2 k$ as Kepler introduced space-time. The formula rests on the relevance **k** applying in order to give the atomic ratio the correct relevance.

In the book dealing with the "sound barrier" I indicated the barriers applying on movement and the limitation that the Roche limit brings to such movement at certain heights in the atmosphere. The movement I limits set by the governing singularity capping movement within its sphere of influence. The first sphere comes about at $\Pi^2 \div 2$ where the limit comes as part of creating a surrounding sphere in which the moving craft establishes individual space-time and the second barrier $\Pi^2 \div 4$ shows the outer limit of the atmosphere. Any object that exceeds these limits will become what became to the cosmic debris that destructs at Tunguska in 1908. The Roche limit of $\Pi^2 \div 4$ came about as the comet / meteor / meteorite/ very large piece of cosmic rock which entered the atmosphere at an excessive speed exceeding $7(3\Pi^2)$ ($\Pi^2 \div 2$) ($\Pi^2 \div 4$) ($5\Pi^0$) = 12618 km / h that is the Roche limit of the Earth.

The Earth relevance came in place and expanded the atomic relevance that liquefied the atoms of the incoming rock and turned the rock into liquid that then vaporised and turned into expanding heat within the atmosphere's confinement and the rest is history. If any thing enters at a velocity more or less or equal to $7(3\Pi^2)$ ($\Pi^2 \div 2$) ($\Pi^2 \div 4$) ($5\Pi^0$) = 12618 km / h or in excess of that limit, the object will turn into liquid. What type of object entered Tunguska is not important because it is not the objects' make up that is of concern but the Roche limit that came into the equation that set the entering object from a solid to a gas that holds the key. This very same phenomenon applies to every spacecraft that enters the atmosphere and in the event where the protective material does not cover the entire spacecraft. The Roche limit vaporised the spacecraft as it did with the space shuttle in 1987 as well as the space shuttle in 200? The protective layer did not cover the entire shuttle and allowed the Roche limit to expand the atomic material forming the shuttle. This had the entire craft vaporise into liquid air that we call flames. However, this also does occur every time a spacecraft enters the Earth's atmosphere and the blanket of liquid heat or flames that covers the craft is the result of the Roche limit coming into effect.

I am not going into detail as to what happens in the atoms because the explanation is comprehensive and I do that in **an Open Letter on Gravity.** There I show what happens inside atom when the Roche limit comes into effect.

In stars with the Roche limit applying the envelope forming Π expands as the lesser star has to adopt the gravity Π^2 of the major star. In this all the atoms within the lesser star has to follow procedure and the atoms start to liquefy as the solid atoms turn into liquid heat. An atom is liquid space-time that spins into a solid structure because the Neutron exceeds the speed of light and the electron speeds heat up to the speed of light. That is the major functions that the atomic particles hold. When the rotation circle widens as result of the interfering of the major star, the speed of spinning slows down and the solid atom's density is brought into question. By not spinning fast enough, the compactness of the atoms in the lesser star expands and overheating takes place, just as it does in the process applying to the nuclear bomb. The atoms in the lesser star goes into a liquid state when those atoms in the lesser star excepts the limit capped by the major star's governing singularity and with the major star's governing singularity in charge of the applying relevance and placing the limit of Π in accordance to the major star's governing singularity, every aspect defining space-time within the lesser star goes array. This brings on the liquefying of the lesser star and by employing the Coanda effect the major star consumes the lesser star as a liquid.

This action I saw happening with the Roche limit then laid the foundation of my theory that everything in the Universe is heat in some or other state of development. There are heat controlled and compressed by rapid motion and there is heat without motion that expands as the controlled heat condenses heat from the expanding movement. The solid atoms consume the liquid space in order to maintain being cold and in that the relevance of density shifts from the liquid or outer space to the growing solids that increase in size. Nothing in the Universe are able to expand because there is nowhere to where such expanding can go and what we see as expanding has to be a shift of heat in density relevancy. I have written many letters on this matter.

The Roche limit is based on movement related to space within time. There are three forms of singularity that I so far have identified and I got around naming two of the three. The one forms as the structure rotates while honouring its axis and that I named the **Governing Singularity.** Then one forms as the object (for instance the Earth) honours the centre of the Sun and that I named the **Controlling singularity.** It is the movement forward which we see as going straight that takes the circling movement all the way around the Sun and brings the total influence of the entire structure to circle the Sun. Then there is the third one which would either be the atoms bringing about the governing singularity moving around the controlling singularity or it would be the governing singularity taking the controlling singularity one step further as the entire solar system moves around the centre of the Milky Way. This movement will be the third movement of singularity applying a reference. I have to come up with some name in order to describe the function.

The Roche limit is about movement and movement is about displacing material in space through time. At a distance of 12 756 km x $\Pi^2 \div 4$ km the lowest speed one could travel and still meet the equilibrium of gravity is the following formula $7(3\Pi^2)$ $(\Pi^2 \div 2)$ $(\Pi^2 \div 4)$ =2523.6 km / h. The value of Π^2 refers to movement at a specific height formed as Π^2 and being $7(3\Pi^2)$ km / h. Then the next level would be achieved at $(\Pi^2 \div 2)$ and the third level will be valid at $(\Pi^2 \div 4)$.

These borders forming the Roche limit are limits taking space into certain heights and depend on density applying. Also the distances applying in km / h are set to standards applying on Earth and only on Earth. It is taken in ratio of the movement the Earth applies as space moves through time.

However the time then applies to the Earth and only to the Earth. One cannot take the kilometre of the Earth and use it on Mars or on the Sun because the space movement going through time on those structures don't remotely apply as it does on Earth. A kilometre of movement on Earth will become (I guess) a billion trillion of Earth kilometres in relation to what would apply to a dwarf star such as the Sun and immeasurably in comparison to what applies when real dynamic stars are involved. The main concern is that atoms expand in relevancy by accepting the relevance of the major star applying and the movement speed up to a point where the atoms in the minor star just simply liquefies. The process whereby this happens is simple to understand once a person looks at the process of movement carefully.

Everyone in general has this idea that moving an object is shifting the entire object from one location to the next location as if one would push a bicycle from here to there. When looking at accidents and the way a car crumbles and crumples in a collision it is clear that the car arrives at the scene of the accident in stages. The car in reality shifts from one point to the other point by relocating every fragment of every atom in breaking down the atom structure and rebuilding it at the next destiny. In this we find the two types holding singularity-movement in charge.

The governing singularity is taking all the movement around a centre and the controlling singularity shifts the object around a controlling structure as the Earth is to the Moon or the Sun is to the Earth. It is the task of the centre governing singularity to keep the atom in tact (T^2) while the movement of the entirety (a^3) relocates by the process of the controlling singularity and (**k**) shifts from one to the next location. Remember the governing singularity of the Earth is in relation of keeping the body of the structure of the Earth intact while the controlling singularity is vested in the Sun and is the orbit the Earth holds around the Sun and therefore the Sun holds the influence of the movement in relation to the controlling singularity. In the event of the major star taking control of the centre gravity of the minor star the controlling singularity is not able to sustain the integrity of the atom while the controlling singularity relocates the entire atom.

From the past
$$k^1 = a^3 \div T^2$$

Going into the present
$$k^0 = a^3 \div T^2 k^{-1}$$

Onto the future
$$k^{-1} = T^2 \div a^3$$

Movement cools heat in order to control expanding. With movement the object moving increases the occupied space-time and by increasing its volumetric space in use, it decreases the heat it holds by spreading the heat over a larger used area. Stars are quantities of heat that reduces the heat found in space by the movement it exerts in space. Outer space is heat expanded to the limit and stars are heat compressed to a point as far as the movement will cool the heat levels at that point in the cosmos. The reason why the top can spin erect with individual movement is that the top gained individual time applying within the influencing sphere of the time applying on Earth but the top has a time apart from the Earth.

I don't care much for the way science defines Universal time because when strictly applied as science defines Universal time then the Universe has time starting in Greenwich which is the main naval base for the British fleet and only a Brit will show that weak mentality as to think that the Universe has time starting in the British Navy headquarters. Brits all over think that the game of Cricket start and ends at Lords and Rugby is only played at Twickenham and British parliament rules the Globe…and time starts at Greenwich where the British fleet is based and built… and all this they believe just because they believe Britannia still rules the waves.

I might even presume that that is why cosmology is in the Newtonian state it is in and that is bad! A Universe spinning that is as much as it is also holding singularity everywhere and in that everything that is spins around a centre holding singularity still within one hypothetical position represents time. It is eternity spinning around infinity where infinity can be any point but also be only one point leaving eternity to be all of the rest of what would form the entirety of entirety.

The movement of the Sun cools outer space to a point where outer space turns from the cosmic gas it is to a cosmic liquid. The Sun is a liquid fridge notwithstanding the overwhelming idea that says the contrary.

In the Universe there is one substance, which is singularity that I call heat but that can just as well be space. The one falls into the group we think of as material and the other falls into a group we think of as liquid. In the centre and forming $\Pi^0\Pi$ we have heat very much controlled by movement representing the solid part and on the other side of movement we have liquid formed as $\Pi\Pi^2$ that holds no specific form.

The two forms I have just mentioned being gas and liquid are the same and the only difference there is, is the state in which elements can find form being one of the two formed by Singularity. It is a choice between the two forms of having liquid/gas and choosing solid elements. Holding elements we find heat forming a component that material could occupy where the space that forms will allow such a compact substance such as material but still it is heat that is forming space being vacant or filled. Space is one side and being compressed it is liquid which is a denser form of space or gas but it then manifests as the heat we know. There is a liquid which is a denser form of gas and then we have solids, which are a denser form of liquid, but in all it is still singularity that is heat in some form. The two forms holding the liquid I called the cosmic liquid and a cosmic gas where the liquid is the more compressed form of the gas and then the third form is also a much more compact form of heat which in that case we then call the state in which the elements are being a state of solids. The form of being liquid or gas or even solids alternates and the changes come about with more or less heat being part of the density factor. It is the intensity of movement bringing about the density that the space holds. The heat forms space when filling a larger area and then being more (when in the form of gas) or less (when in the form of liquid) or absent (when in the form of solids) establishes and reforms the state in which the elements cluster together or we have heat in a specific formation between atoms. Nevertheless liquid and gas is ranging in heat levels being compressed and dense or expanded and less dense where heat is the opposite of space but also is the very same thing that came about before the Big Bang event took place and represents a period at the time the cosmos was still forming the second form other than atomic solid elements.

See the fluid push out of a bowl of liquid we think of as the Sun. That we see that is spilling both sides as it falls back into the Sun is not gas but it is liquid. The Prominence is liquid that falls back into a bowl of liquid and the Sun is one spinning sphere filled with liquid. The inside of the Sun is not gas but it is fluid. Stars are material floating in liquid and in that we have either material or fluid, but there is nothing else found in the cosmos than the two variations formed by singularity that is holding singularity in one or the other form.

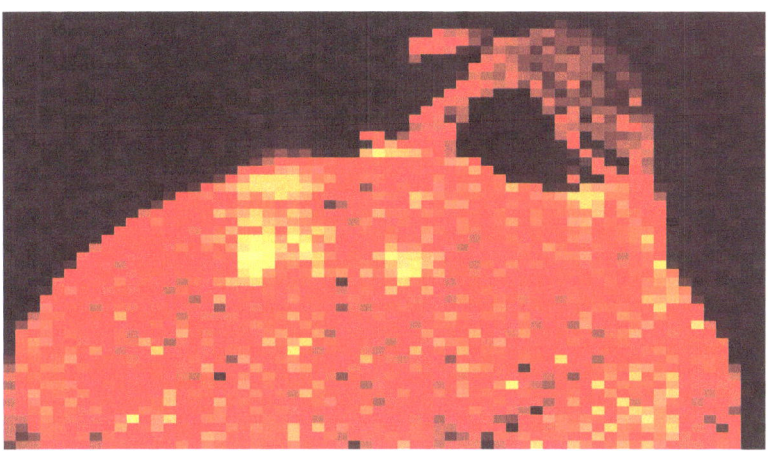

Every one knows that a gas is one dimension HOTTER than Liquid as liquid again is one dimension HOTTER than being solid. The heat surrounding elements is not a match that fit all and is a separate issue in every case with every element. The heat /space /material relation gives elements their characteristics in form they are recognised by. Such characteristics changes as the heat /space/ material relation changes and alters the presumed characteristics

Hydrogen 1	melts at -259^0 C,	boils at -252^0 C,
Helium 2	melts at -269^0 C	boils at $-268,9^0$ C
LITHIUM 3	melts 180^0 C	boils at 1300^0
BERYLLIUM 4	melts at 1287^0C	boils at 2770^0C
BORON 5	melts at 2030^0 C	boils 2550^0 C
Carbon 6	melts at 804^0C	boils at 3470^0 C
Nitrogen 7	melts at -210^0C	boils at -195.8^0 C
Oxygen 8	melts at -218.8^0C	boils at -183^0 C
Fluorine 9	melts at -219.6^0 C	boils at -188.2^0 C
Neon10	melts at -248.59^0 C	boils at -246^0 C

Melts (meaning that the element becomes a liquid) at -259^0 C. **Hydrogen 1** boils (meaning that the element then becomes a gas) at -252^0 C**, This does not concern the basic atomic element but only applies to the space on the outside of the atom**

By the way I leave just a thought: This could just as well be the "matter and the antimatter" that science is searching for when science claims there was a partition that came about at some point of cosmic development and as they put it the one part started devouring the other part in the matter vs. anti-matter fighting, but this is a solution I offer according to my opinion. Some particles became matter and others became antimatter or plasma or heat or just what you wish to call the by-product that came about from the friction that brought about the heat that led to the expanding of the space. Those forming units we call elements had established characteristics relating to space-time where each one holds an individual identity according to the number of protons the element has in one cluster unit. There are elements. The elements presume in the role of being solid and form as units the solidity of solid materials. Then there are liquids. Amongst those elements mentioned, we regard some of them as being natural "liquids" and others being natural "gasses", however they are solids notwithstanding what our culture call them. There are those we consider to be mainly gas or liquids and only then there are the state of being "frozen gas" (as we regard outer space to be) but such presuming underlines our mistaken culture we have.

In our human culture we swapped the definition of hot and cols around. When something is hot that something expands and what can be more expanded than outer space is. When something is cold that something becomes dense as it contracts and what can be more contracted than what solids inside an atom is. This is a nature law that humans in science missed and that could not be ignored. Within the Solar system what could be more contracted and denser than the atmosphere around the Sun. When that space becomes dense then to our view we have, that space becomes hot because we feel the heat. This is a long and arduous argument I could never complete in a book such as this but I do argue this opinion of mine to lengths in other books, which I have already written. A star is a compacting devise that cools the overheating outer space and in that controls the overheating by applying movement where there is a total lack in movement. In **THE VERACITY OF GRAVIY** I argue this point in much detail.

We can see the star that expanded due to overheating and in the process it shows the star holds liquid containing material. If the star was gas that expanded it would be invisible just as the black part surrounding the

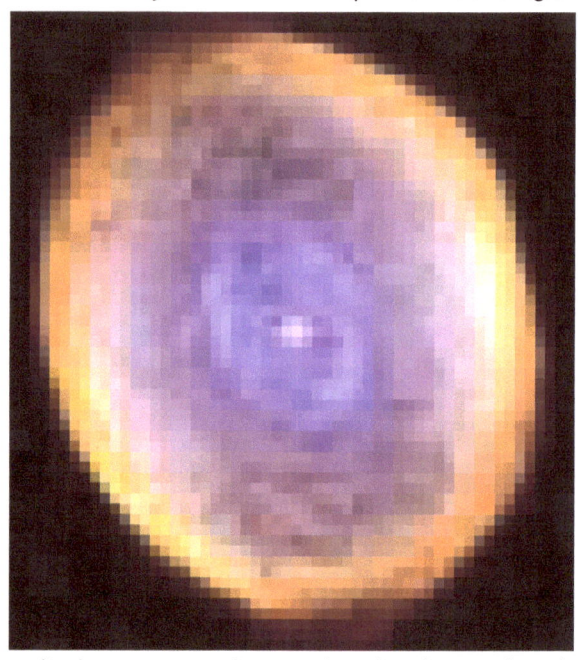

star is, but to be able to transmit heat in the form of light it has to be solid enough to be liquid because light is the most solid any cosmic liquid can get. If the star is liquid on the inside holding material, and the liquid evaporates when coming into contact with outer space such as the case is when the prominence squirts out into outer space, then outer space is the hottest, notwithstanding what ever boundaries and values we humans attach to the dimension. When something is compact it freezes by losing space. When something is hot it expands by gaining space. That is a law of physics no one can deny. When dealing with cosmology our human standards have to change to accommodate the rules laid down by the cosmos and not apply our personal interpretation of the cosmos to suit our rules of hot and cold, big and small, near and far. In the case of the Super Nova, something prevented the liquid turning into gas, therefore overheating before the event where the Super Nova took place. The liquid is cooled heat frozen by movement to form a liquid and the movement cooled the heat to become a cosmic lollypop.

That movement or gravity spinning the star, which is what gravity is, prevents the overheating by turning the layers within the star into frozen identities and the movement cools that star and prevents overheating, therefore it became a liquid outside the star. This turns the star into a miniature galactica, sustaining billions of individual points that represent singularity, because the governing singularity did not destroy, but the singularity of every nature is still in support of one another. From this picture (and others of Super Nova) one can learn a lot, if one is truly interested in applying cosmic law to the picture and not some human response to what we think would apply to an earth-like star that holds gas as an ingredient. A Super Nova is not a star holding gravity that has gone mad because gravity has no intellect to lose. A Super Nova is simply a star that was turning to slow in some or all of its layers and by spinning too slow there was insufficient cooling whereby the layers overheated and expanded. This is the Roche limit that allows the minor star to simply overheat into liquid.

Outer space is heat that overheated to such an extent it is still over heating because it is still increasing volumetric space. Outer space is heat in such a heated state it can expand no more but by the margin time allows it to do so. This statement explaining the process in which this happens is far too complex to discuss in this book and I do so in others that allow much more conversation. But with a slight bit of intellect the feasibility of my arguments are much more acceptable than to think gravity is magic that can go mad in Super Novas.

This gives a real explaining to the cosmos and the explaining shows how simple and transparent the cosmos can be. The way science thinks of the cosmos and puts nothing into any equation that renders gravity the possibility of becoming magic and grants the cosmos magical powers is very loony in the least. By not being able to say what forms gravity, then one grants gravity magical powers and that leaves the cosmos in a fairytale with magic running long ad short. By excluding nothing from the equation space becomes something bringing in a value lying inside the realms of the infinite that must form singularity. Applying this logic to the Lagrangian system as well as the other three phenomena and interpreting that information to the law of Pythagoras a clear pattern comes about.

This takes us to the Coanda effect.

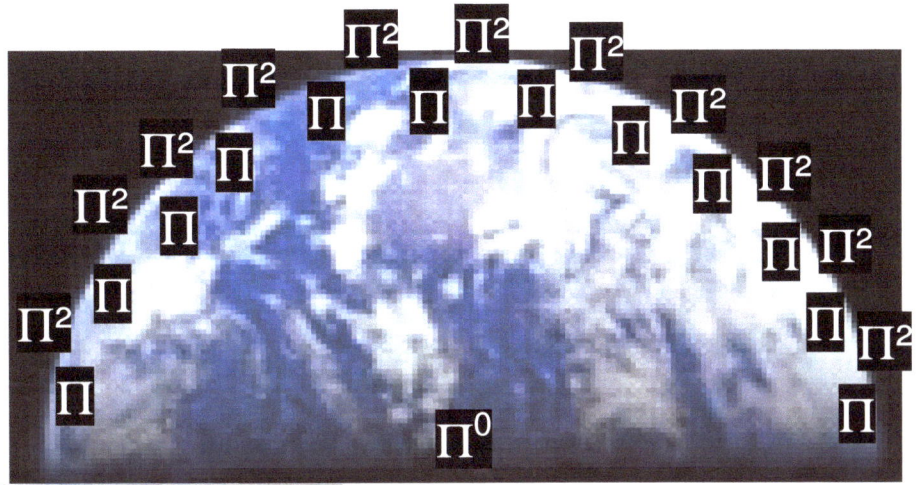

Again I have to press the thought that it is singularity determining space-time To think that matter can be solid liquid or gas in incorrect It is the condition of the space-time derived from singularity that places the form and conditions valuing the form of the elements into a balance between solids and much a solid as gold can be a gas. Being solid or not depends on a ratio that exists between cosmic liquid and cosmic solids

liquid. Hydrogen can be as

By denouncing nothing brings about that:

Matter is singularity that turned solid by movement exceeding the speed of light.

Heat is singularity holding a density in which it is liquid that turned singularity to a cosmic fluid

Space is singularity formed the cosmic fluid that turned to a cosmic gas.

All elements are solids that are parted by a substance. The substance might be tiny in quantity, which then allows the atoms to be tightly packed, and this is a hard solid. When the substance parting atoms are more in ratio it forms a fluid and when it is much more in quantifiable volume that the solids are it forms a gas because it is the relation that requires the state of position the substance form in relation to the rest of the cosmos. Hydrogen is as much a solid as gold is a gas because it involves the unoccupied heat surrounding the element to form the state in which the form is that which represents the substance we see as material. This is what becomes clear when investigating an aircraft going past the sound barrier. It is the result of the heat that the structure of the craft builds up when dynamics change in the phase of changing singularity. In the structure one can note from the cracks in the body of the craft.

This the result of the surrounding heat amplifying, as the heat becomes space when the craft slows down in landing and becomes liquid heat again when the craft exceeds the sound barrier and the surrounding air heats again. The cooling and heating process happens in the air when the space parting the atoms forms cracks in the structure of the body of the craft as the heat and cooling thereof becomes permanent space or body cracks after it was heated too many times with too many heat sessions that left scars and traces of the process that applies. It proves that there are dimensional implications all around the body structure of the craft when flying and then landing once more and that the dimensions that changes are valid. The process of heating is the result of the Roche limit applying but it is because through the Coanda effect gravity is pushed to new limits.

Stars only become stars when the stars start to spin and develop its individual time distinguish it from the geodesic, meaning the stars interior remains solid allowing what is beyond the outer edges to become a relevant liquid when this relevancy is comparing to the geodesic of outer space and having outer space going onto a form of relative gas. When looking at the Earth we can see shows how the "atmosphere" of the Earth carries the value of Π^2 in relation to form the liquid as the surface forms the solid being Π. This is why mass becomes a factor of sorts.

The body standing on the surface holds a connection with all the other points serving the governing singularity and becomes part of the line formed between the governing singularity and the point holding Π and since it is part of the line and forms the eventual Π it forms a factor human can use in a measure of calculation. It is the differentiation brought about by movement that is distinguishing the solid space from the liquid cosmos by turning the Earth movement into a solid that puts space, which stands, still in a position of being a relative liquid. The liquid forms part of the rest of the cosmos being a liquid and it is including everything filling space not within the boundaries of the solid Earth.

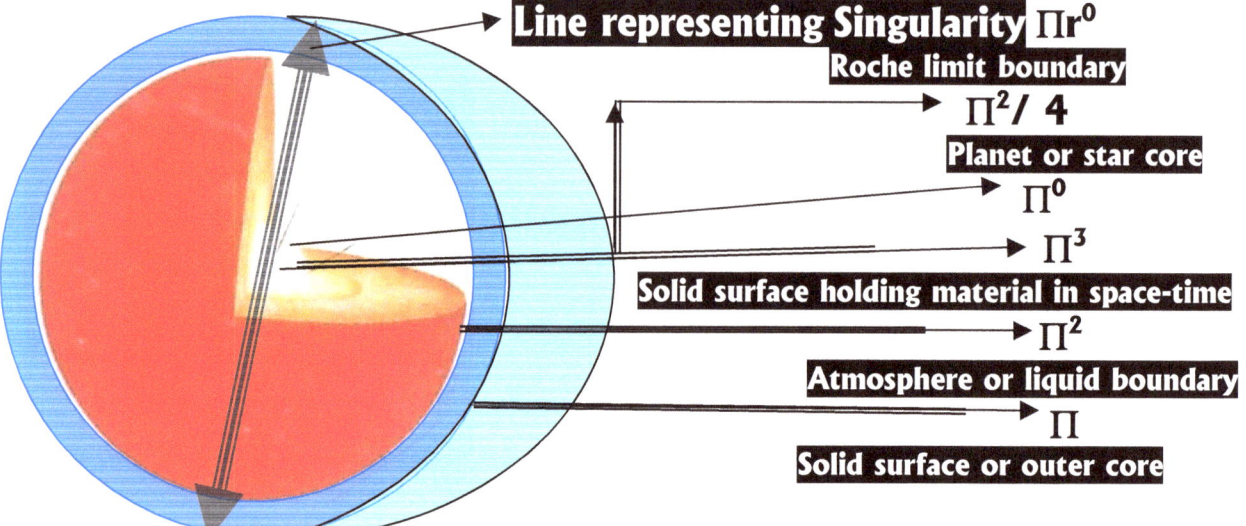

Line representing Singularity Πr^0

Roche limit boundary

$\Pi^2 / 4$

Planet or star core

Π^0

Π^3

Solid surface holding material in space-time

Π^2

Atmosphere or liquid boundary

Π

Solid surface or outer core

These boundaries are not specifics but relations to certain limits set from the position of singularity outwards. Te boundaries are set in relation to the Roche limit applying and setting such limits.

In the centre of material is singularity forming infinity. On the outside of material is singularity forming eternity.

Between singularity in infinity and singularity in eternity movement introduces a divide we humans think of as matter or material. That in principle constitutes the Universe which by right is formed by singularity and singularity does not exist but for relevancy applying set by movement and the rest or individual elements of insignificance because size has no meaning in the cosmos.

Every atom is a small microscopic pump that draws heat through the space between all the atoms and mass is the restriction that the flow presents.

The star is a centrifugal pump that pumps heat as a coolant towards the centre of the star and this is dome by the protons in every atom. The protons expand and contracts and in that way it deliberately "sucks" heat from the outside through the electron going and flowing with the liquid neutron towards the proton that then delivers the heat to singularity in order to maintain singularity in temperature.

When a star starts to turn by forming an axis that forms singularity the solid part separates from the liquids where the spinning atoms form the solid and the rest becomes liquid having a fluid atmosphere. Only then can it start performing its duties as a star by converting space (gas) to fluid (heat) which then turns to matter and matter back to singularity. This is achieved by combining protons to denser proton clusters and in that manner serving singularity in a more supportive manner. If the Sun or any other star were using gas, it would not have the ability to generate the means of sustaining fusion because "gas" can't compress well enough to compress solids into bigger chunks of solids. Only heat as liquid in the stars would have the ability to compress material. This is how stars form and that is the basis for galactica being in place.

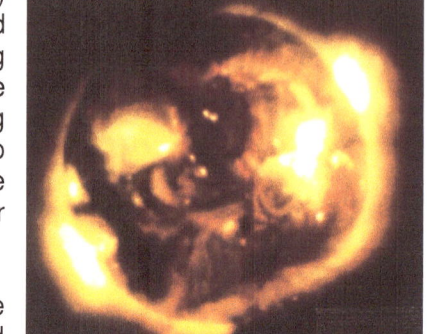

In **the picture of the Sun** we find not withstanding whatever name we attach to the **red liquid substance flowing from the Sun** into space and back to the Sun, **that liquid is heat** in a very direct form. **If outer space was the coldest place in the solar**

system the heat **should** immediately **escape to outer space** and **not return to the Sun** as it clearly does because if outer space was the coldest place it will freeze everything and have the lot flow towards the shrinking cold…that it does not so therefore the Sun has to be the coldest because it is shrinking outer space having outer space flow towards the Sun. If **outer space were colder the heat would not return to the sun**. Kepler's figures prove that space within the entire solar system is flowing towards the Sun and that forms the gravity the Sun provides the planets as to keep the planets circling the Sun.

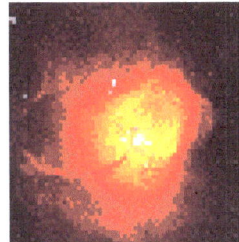

Stars can and stars do **overheat**, sometimes and the **Polar Regions** where **the Titius Bode matter-to-matter applies** holding the square matter (7+7) in relation to the square of space (10) and **other times** in a double relation to the **square of space** 10 to that of matter in a half square (7 /10 or 7/ 10). When saying the above, as I just said, one has to differentiate between heat and overheating because a star represents the coldest space in the Universe and not the hottest space as Newtonian wisdom wish to interpret when using Newtonian standards to read into laws applying in the cosmos. **Heat and cold are relevant dynamics** forming **in appreciation of singularity.**

The Sun is the coldest place in the solar system and that is fact notwithstanding what man tries to tell the cosmos how it should be because that view may fit mans perception. Looking at evidence the Sun provides, it contradict everything science wishes to believe about cold and hot. Science wish to see the cosmos through the eyes of what fits the needs sustaining life on Earth and what benefits in maintaining the surroundings as to support the security of life. Man has this novel idea to look for conditions as one find on Earth whereas life has no part in the cosmos except for being located on parts of the speck of dust we call Earth.

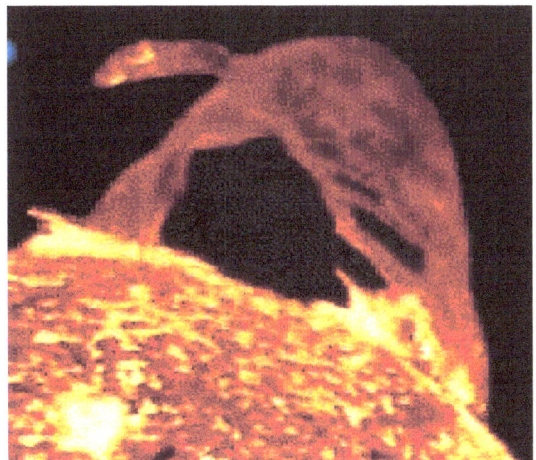

No person with normal eyesight can deny the fact that the prominence squirting from the surface of the Sun is a liquid substance or a fluid. This fluid can only be the purist liquid one may find anywhere. It is cosmic gas that chilled down to form cosmic heat in the face of the cosmic material within the Sun. When a substance such as liquid squirts into space in the manner that it happens within the Sun, this could only be as a result of high temperature differences occurring within the Sun where high volumes of heat makes contact with relative low temperature material within the Sun and the differences causes this violent expanding. If it is liquid squirting from the Sun it can only be liquid the squirting comes from and that makes the Sun not a hydrogen gas bowl but a liquid pool filled with the as pure a form of fluid any place in the cosmos.

Looking at the cosmos in an impartial manner we find that life is not the pinnacle of the cosmos and to the cosmos life doesn't even exist. In terms of the cosmos, life is an afterthought added on and in term of the entire cosmos as far as all evidence goes life is only located on one tiny spot we named Earth with no cosmic significance. If we wish to study the cosmos we have to look for the evidence that supports another view other than holding life as the panicle. Every aspect in **the cosmos is the very opposite of what science believe** it is. The Sun is **not a ball of gas but** a **giant sea of liquid**, frozen **without any** form of **gas or air** in the interior. Having a liquid interior **the Sun** has **no pressure** but has the **very opposite of pressure** to which there is yet no name given. **The liquid comes from singularity freezing** space-time within the atmosphere of **the Sun**, and such is the case with all stars still in the shining phase. **Stars more developed than the Sun is frozen solid causing fusion.**

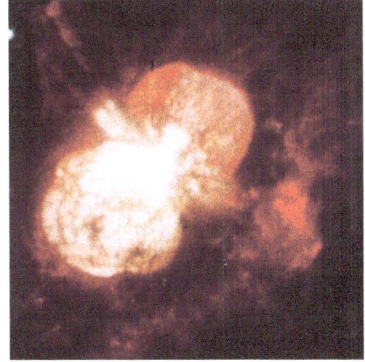

Since motion in stars forms the condensing of cosmic space to become a liquid cosmic fluid, the movement causes a cosmic gas to go on to form cosmic liquid heat. That is condensation resulting from material spinning. The top spinning condenses the air around the top and that is what keeps the top erect. The spinning causes the differentiation brought about by the movement of material. The density of liquid heat results in much higher thrust and through the drive obtained from this liquid thrust this forms a much higher transfer of power than what a term such as mere pressure would suggest. When an explosion demolishes matter, no force in the cosmos will stop the accompanying destruction. This we find in nuclear destruction of atomic structures such as we have in Hiroshima and Nagasaki and the Bikini Island atomic test as well as in a limit way what we find present at Chernobyl.

A Super Nova is the same process but going on a scale man still has to invent words to describe the difference. The reaction starts because there is a massive unbalance in the relevancy of space occupied to heat bounded by matter to specific space occupied. The heat levels in the star surges because the movement of the spinning star is not sufficient to cool the star structure and the atoms are not equipped to generate a strong enough governing singularity to produce the gravity (or movement) to maintain the required heat levels and thus produce the required cooling through movement. A sudden super abundance of heat coming available places space occupied in a disadvantage to space available since all the available heat became available space through the process we refer to as an explosion. This process has all to do with motion not creating cooling conditions that will freeze space into a solid condensed liquid state.

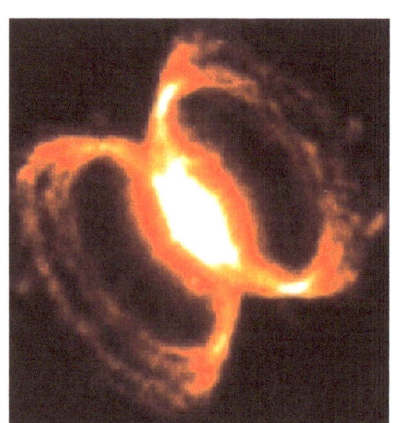

This exploding star is a picture of the release of heat that was previously held within the star. If the star were anything other that one bowl of liquid heat, this picture would not be what it is. It is clear that the released liquid heat tries to maintain a form of Π as the liquid through gravity tries to maintain a connection with the governing singularity. A picture such as this shows that a star is concentrated heat that can overheat and the only way it can overheat is when it's cooling goes array. The cooling comes from a process of movement and the more rapid the movement is the more it freezes everything onto singularity and then the colder and more compact the star is. Gravity is the cooling of space by implicating movement to concentrate heat that surrounds material. All elements forming matter in as much as being the heat concentrated by movement exceeding the speed of light when then forming an atom can be as much a liquid as it is a gas and a solid. With little heat in the ratio all elements are a solid and with overwhelming heat in the ratio all elements forms a gas. In the cosmos there is no hot as there is no cold. It's about storing energy in space or in heat, which is another cosmic equal being opposing similarities.

Hot and cold are relevancies brought about by singularity valuating space-time and during the Big Bang the Universe was freezing cold at three billion trillion zillion degrees C. The measure of heat is not important because if the one point forming the heated value was exceptionally hot, then the other point forming the cold limit was exceptionally cold. There has to be a zero to have heat finding a measured relevancy between two points forming the limit to hot and the limit to cold. It is the relation matter has with heat that provides the form the particle has at that moment. Say during the Big Bang space was 10^{37} degrees Centigrade in space then he points holding material were zero but if the points holding material were 10^{37} degrees Centigrade, then space was just about minus

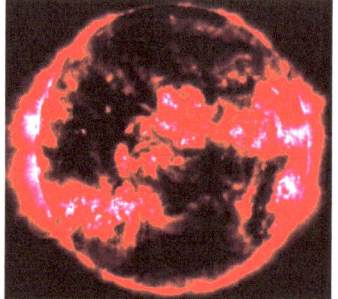

zero. If there is a scale, then there has to be limits to provide the scale legitimacy. The increasing or decreasing of the heat levels will alter the form of the element. Therefore all elements forming matter is as much a liquid in composition or not than it is a solid or a gas. It is the space surrounding the atom that hold the non solid base which provides the form the atom finds its relativity to the rest of the atoms it share space with. Hydrogen is as much a solid as tungsten is a gas and the form depends on the ratio of heat in relation to the solid within the space holding the matter. If the argument in reply is that it is the gravity pulling the heat back to the Sun, then that confirms my theory that gravity is all about reducing space and eventually confining heat to such a small area that gravity is about collecting heat onto matter and with that statement it then declares that outer space is being the hottest and therefore the most expanded place. It is the concentration of heat in space being relevant to form and that come about as cooling reduces heat where any increase in heat will bring about an increase in expanding. When a star is overheating then a star turns its liquid to gas whereby it merely transforms its interior to a relevancy it has from the pre- to post- Big Bang era.

We humans on Earth think that hydrogen is a liquid at -259^{0} C but that only apply to the Earth. Looking at pictures taken from the Sun we see in such pictures that it clearly shows the heat in a liquid flowing from the Sun and back to the Sun. In the Sun the hydrogen holds enormous quantities of heat in a liquid at a temperature of 6500^{0} C. When a star has its singularity secured, the star is bitterly cold because it has heat in a liquid form flowing back to the point of singularity although we may regard the star to be rather on the hot side, however that observation serves human interpretation and is very bias to life. The cosmic truth is that the Sun (for instance) freezes hydrogen to a liquid form at 6500^{0} C. The value of 6500^{0} in terms of hydrogen being in one or another state in terms of

human concepts is meaningless because the movement of the Sun and the cooling of conditions on the Sun is so much more influential than what applies on Earth that the drawing of comparisons are just proof we humans are without understanding. If hydrogen remains a liquid at 6500 ^{0} C, just think how cold it must be as the star's interior approaches the point of singularity. Therefore fusing protons comes from cold and not from heat or pressure and in that way the process becomes sensible. By allowing the singularity to overheat all the atoms in the star must overheat and then the star overheats and heat within the star flows from singularity to outer space freely. In such an event outer space is then colder than the star because the heat releases to outer space with no intention of returning whereas under normal conditions in the Sun it returns as soon as it leaves. There are two ways to reduce heat; one is to bring about expanding space, as the photographs clearly show. The second one is where heat will reduce when an object is in motion by spin. When withholding motion or critically retarding motion then the reduction of movement will bring about that matter will overheat. Gravity is the motion of unoccupied space through the dimensional transformation to occupied space. This comes about as the star moves and therefore duplicates its position in space more rapidly and with that it will spread or distribute the heat over a larger area and in that bring about a drop in heat levels in the entire area.

Motion is cooling which is pace reduction and then to counter this statement not moving of space therefore is the anti-, the opposite, the negative and therefore expanding of space being in contrast to cooling that is the concentration of heat. Condensation of heat is a process of cooling space where this cooling process will rather seem as exaggerating the heat levels by heating the space and heating will lead to expanding which will to us seem as if it is cooling because the higher heat levels are also spread over a bigger area holding space. With singularity overheating the expansion of the singularity drives heat to form more space, creating space to compensate for the levels of overheating. **That is a natural phenomenon**. The only reason why **heat will** rather **flow back** to the star than **escape to outer** space once the star released it into outer space is **if outer space presents more heat than does the star,** because **heat always flows from hot to cold** no matter what influences may arise. **Outer space must hold more heat than does the star but the accumulation of space in relation to heat makes it seem colder bringing expanding of heat to become space. <u>Space and heat directly relates being the one form of the other</u>**.

The cosmos is all about **converting space to heat** which we see **as gravity** and **returning heat to space** as a **control mechanism** always **keeping** a very delicate **balance** which we see as **a star shining or being normal.** The purpose of the converting of space to heat is to supply the core a dose of heat to cool the point holding singularity where singularity is with heat. **It turns space to heat** sustaining matter but sometimes singularity overheats and then matter converts to heat allowing heat to convert to space. That we call many names amongst others exploding into a Super Nova.

Whatever the names used is less important because the **process rests on space and heat interacting to form energy** but I am reluctant to use the word energy since energy is used for chocolate as much as it is used to describe petrol as much as it is used to describe life's vigour. The concept behind the use of energy in terminology is like a whore. It is everybody's wife while it actually belongs to no one. The release of heat to form space was what **the Big Bang** was and **the Hubble Constant** is all about where space not holding **matter converts heat to space.** I show that **space and heat is the very same thing** and there **is no such a thing as pressure** but releasing **heat produces space** and **concentrating heat reduces space** with the two interacting on singularity demand setting time to space with time being the spin or motion of heat in space. **Heat and space form the second singularity** caused by the **fragmenting of singularity to compensate overheating during the pre-** Big Bang matter forming era. That is what we see as **light and space,** which again is the **same thing and is fragmented** to hold points **serving singularity forming radiation and heat, where the star re-transfers heat back to space due to an overload.**

Looking at a star overheating it is obvious how singularity disconnects the control it has and the space once tightly packing the star by spinning movement losing control of the movement and the interior of the star then breaks into heat. By demolishing singularity it means $\Pi^0\Pi$ and $\Pi\Pi^2$ demolishes the very point holding singularity atΠ. There is still a centre but the control the centre had is lost as the liquid forming fluid heat spills into outer space. However, take notice of the fact that the 7+7 /10 ratio of the Titius Bode law is still in effect.

The fact that stars overheat is never mentioned and in that the question never asked is why would stars overheat? We can blame pressure, but pressure would not bring about a star disintegrating from the centre, as the star depicted here clearly does. A burst from pressure should blow the sides out.

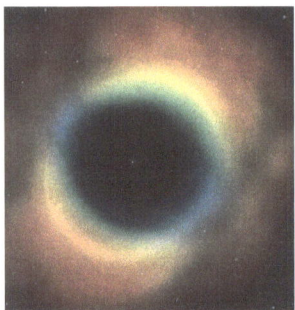

Stars we call Super Nova has blowouts. That man knows since before writing began, but since of late this phenomenon becomes more and more seemingly misunderstood. If stars blow as stars should and as we can clearly see from the picture, then the explosion happening to the star just above this Super Nova surely comes about from other principles. It is very obvious the two occurrences are not a result of the same basic method the Universe uses in destroying stars. When looking carefully to what happened the centre holding the once governing singularity is still present but the heat became space as the gravity became too slow to keep the integrity of the star structurally intact.

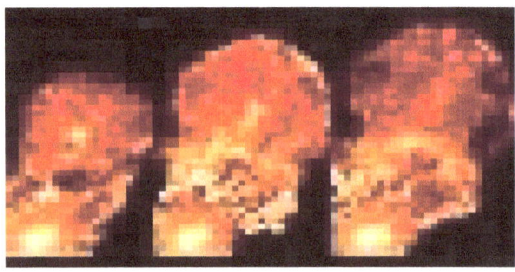

When heat surges and becomes too high, it turns into space. That process we call an explosion but are just heat reforming value as it reshapes concentrated heat into an equal measure in space. It is frequently seen, yet it is never acknowledged by science. When heat reduces by cooling, it relinquishes space in the producing of more concentrated heat, and this process we see is cooling. **On the other hand it is true that the reduction in controlled heat brings about increase in heat that manifests as space.**

Whatever the terms are used to **describe a process there must be a recognising of the inter relation between heat and space where the reducing of the one will lead to the increase of the other.**

The star does not apply pressure to bring about fusion, it freezes (actually it freezes time but that concept I argue in a far more specific book) and by getting time to stand almost still, the elements compress into fusion. This it does by applying millions and even billions of degrees Celsius. It is our conception of hot and cold bringing total confusion about the principles of cosmology.

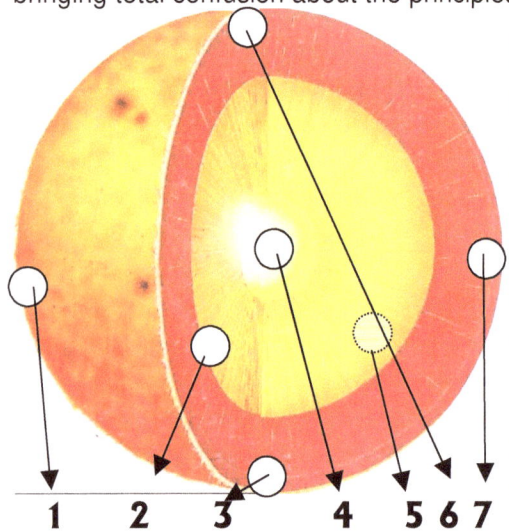

Elements do not determine form because any element can be a solid as much as it can be a liquid or a gas. It is the state of heat between the elements determining the state or form and that comes from singularity governing space-time. The layers in a star stand under separate conditions with space-time. In the sphere Π^0 in singularity holds equal Π in six specific points becoming a total of seven Π and no radius. When removing Π^0 Π in singularity the circle loses some and sometimes even all of its integrity. Where the singularity no longer control space, Π compensates by relinquishing space to r. where Π removed its value r becomes the square and the circle then becomes $\Pi r^0 = 10\Pi$. In the normal tongue we call it the atmosphere.

1   2   3   4   5 6 7

Every star is a Universe holding principles applying only to that star and standing on Earth while viewing the star it is impossible to fathom the values applying within the star because as we stand on Earth we can only value while being under the control of the governing singularity as we experience it apply on Earth. The fact that it can freeze heat to liquid surrounding hydrogen while holding a temperature of 6500^0 C should be an indication it is not what we seem to acknowledge as normal. The Sun is freezing hydrogen to a dense liquid at 6500^0 while space is boiling and over spilling into more space (expanding through overheating according to the Hubble Constant) at -273^0.

Science academics have to review there thoughts on relevancies because what seems to be hot is cold under certain circumstances and what seems to be cold to a point of freezing is boiling hot. The only constant applying is that there is no constant ever applying anywhere. There are no standard issue and fit all through out the Universe. Every point holding singularity attaches different criteria to borders controlling the space-time with in it rule. What fits humans on Earth does not even suit conditions everywhere on Earth let alone conditions applying on the Moon, yet science can't appreciate that Mars applies very different standards to that of every structure and every structure is a cosmos on its own turf, supplying its own turf.

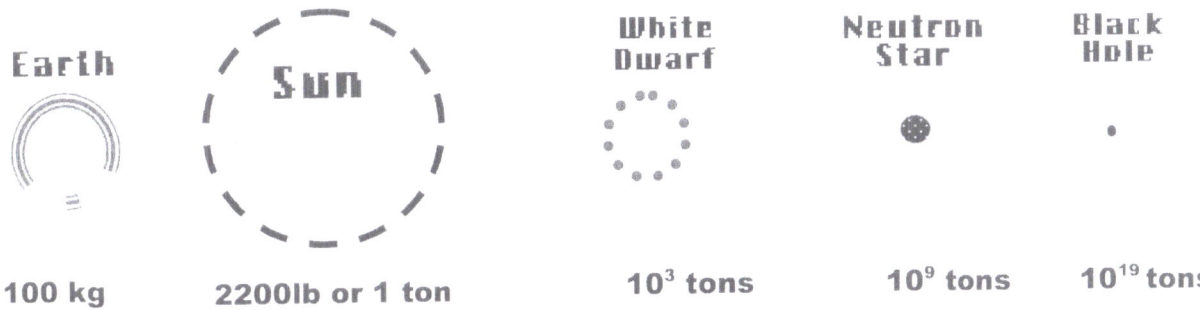

Earth	Sun	White Dwarf	Neutron Star	Black Hole
100 kg	2200lb or 1 ton	10^3 tons	10^9 tons	10^{19} tons

Every structure in the Universe applies $(\Pi^2+\Pi^2)(\Pi^2\Pi)(3) = 1836$ as the atomic relevancy but the range in difference are valid in a different way ranging from a red dwarf to a Black Hole. It is not the specifics that are of importance because the specifics change considerably because this is most apparent when taking into account that hydrogen remains in a fluid-like frozen state at 6500^0 C on the outskirts of the Sun and therefore it is obvious we have to look at other clues to give some indication of what is in process. On Earth in the time we have as a duration we find hydrogen freezing at minus 269 ^{0}C as where it freezes on the Sun at 6500^0 C, which implicate the reduction of space to an enormous increase in time duration.

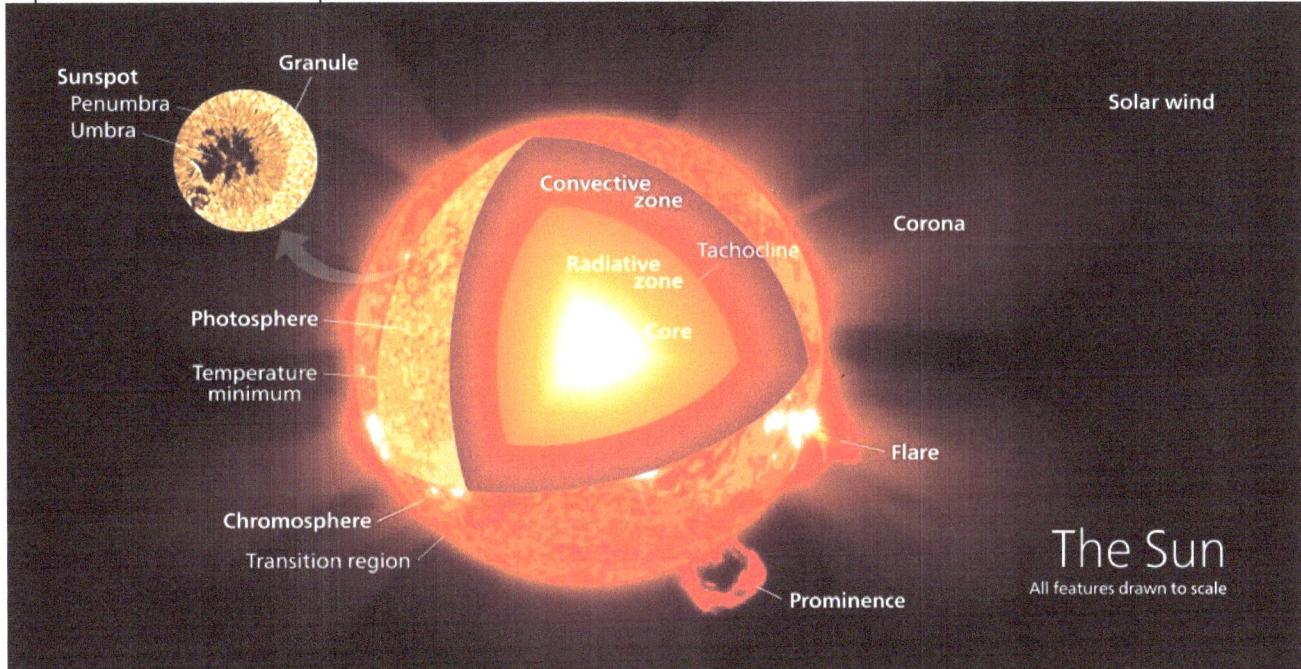

The Sun
All features drawn to scale

In conditions on Earth the rotating velocity of the electron is 3 X 10^5 km / sec. With conditions being that different it can not nearly be the same in the Sun. As space reduces time increases and this is a fact that is a constant. By having the space reduced to such an extent that it matches near Big Bang relevancies (a period where heat flowed like water and which is the very same conditions we find within the Sun) the time would apply accordingly. We also know that relevancies is all about conditions showing similarities under variables and therefore the space and heat component may seem altogether incompatible but is almost the same given the singularity presence within the Sun and comparing that to the Earth.

What is applying to stars inside the galactica centre is applying to particles inside the Sun.

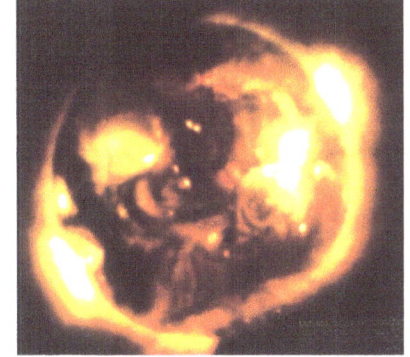

Science sees the nuclear reaction but do not recognise and therefore do not admit that the nuclear reaction is three different phases. One can see that the atom is "canned heat" bottled by spin into a canister. If you break the canister by letting the spin reduce, the heat escapes by going liquid and the liquid turns into gas buy forming gas. That stuff coming out of the atom when exploding reminds very strongly of the liquid coming out of the Sun but that doesn't make the Sun one big atom bomb because then the Sun would reduce by losing its gravity-supplying material. It just says the Sun is what makes the atom container containing what the atoms contain as heat "canned". The Sun helps the atoms within the Sun puts heat into the already existing container by using the four cosmic principles. At the beginning of the nuclear explosion process all the heat is solid, placed in a container by nature and the container has a human name called the atom nucleus. In the atomic explosion there are three ingredients that are distinctly apart. When the solid melts down, it becomes a fluid. The fluid we gave the name of light. There is not enough space to explain the detail of the argument, but light is not a gas, it is a fluid. The first step of the nuclear explosion is converting the solid to liquid. In the liquid state the star does not overheat. The overheating becomes part of the second phase. That phase involves the turning of the heat-fluid to a heat-gas we call space. Space is heat overheated creating space, as heat is space concentrated creating a fluid or liquid not yet correctly named.

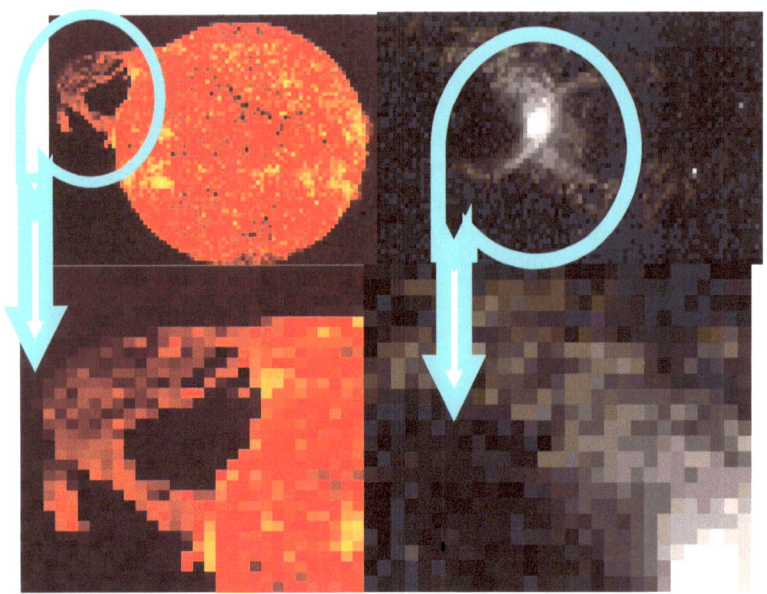

In the pictures to the right the apparency is more than obviouse that the overheating of the star at the core is heating the interior ro a level where is hotter than it was when mvement controled the structures integrety and a blow out into outer space took place and that brings about that the heat will flow to a colder region. In the one case the (Sun) the star is overheating at the edge and blowsw liquid in a squirt into outer space. There is an obviouse difference between the Super Nova that the governing singularity could no longer protect its integrity and the difference of a star NOT overheating being "normal' with liquid pouring from it and then becomes a gas that evaporates and condenses back to Sun . This is more the result of compressing the heat into the Sun and it hits the cold within the Sun. Gravity is the reducing of heat

After establishing the reference point to either singularity reduction or space-time enhancing through allowing matter to grow, the Titius Bode law applies, which I have explained in the pages preceding this page. Total annihilation and destruction of the singularity in one object may result in the object fragmenting to smaller parts where each part will still hold singularity, affected by less matter claiming space.

If mass is responsible for producing gravity and if gravity is responsible for keeping the Universe in check, then mass has to be one part liquid and the other part solid and if not then Newton had it wrong all this time. The cosmos is about liquids interacting with solids and that is what all four phenomena prove. Science can go on to deny this and remain stubbornly witless in their concept as to what goes on in cosmology, or mainstream science can reach a point where they admit that Newton made errors and Newton explained everything in physics, but should be kept out of astrophysics in total.

Every view of the Universe tells a story of liquid being contracted by solids and when movement can't keep the lot together by enlisting sufficient gravity, the lot goes array as liquid expand again and release back into the heated outer space.

When starting to explain the Coanda effect I wish to start again with explaining the forms coming from the divide brought about by $\Pi^0\Pi$ going on to Π^2. The Coanda effect holds its value in the differences there is between singularity represented by liquid and singularity represented by solids.
What ever is in the cosmos a visible or otherwise hold it's meaning

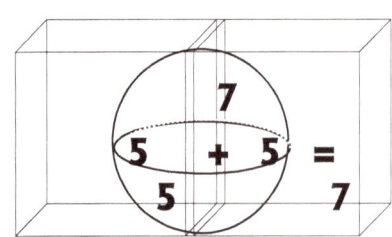

The TITIUS BODE

Principle Inside the sphere
7 / 10
5   = 7 / 10

Space-time is a four dimensional position of the Universe where the position of an object is specified by three coordinates in space and one position in time.

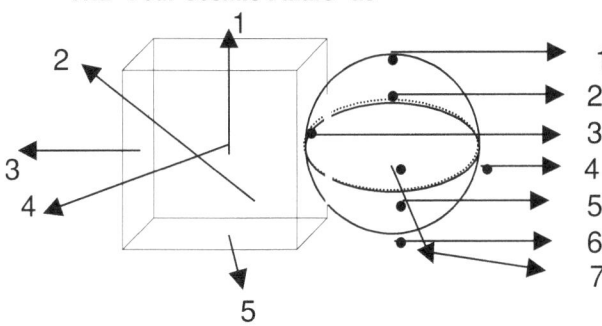

As the meeting of r points to a very distinct different r in direction such a point of meeting opposes the other points in meeting and will lead to destruction of the form Π in any the event of any value changes by Π changing Π^2 and r.

By coming into contact with the sphere the cube loses on dimension to the seven dimensions dominating six bringing about that the cube then has 5 sides to the seven of the cube. That is the Lagrangian system with five cosmic atoms holding relevancy to the centre cosmic atom where the centre cosmic atom stands in for seven and the orbiting cosmic atoms standing in for five positions in space. There is a more explicate explanation about this somewhere else in this book.

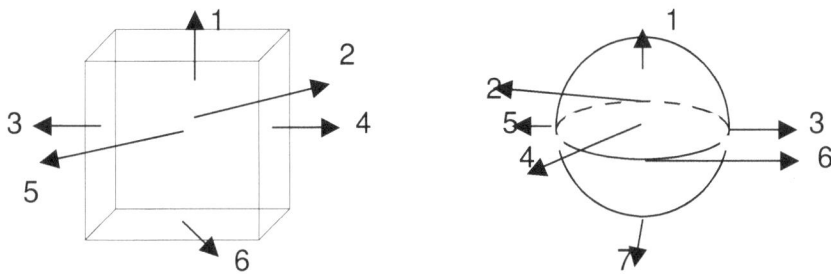

In the cosmic sphere there is no radius but only the extending 1^0 to 1^1 or of Πr^0 running from the centre Π^0 in six opposing directions relating to one another by the square of Π but remaining Π because of the unity the matter holds in relating to space. In every sphere there then are the six $\Pi^0\Pi$ relating in precise dimensional and positional equality to the centre Π^0 as well as to one another by 90^0 and 180^0 implicating the dimensional positioning. Therefore the sphere holds $_7\Pi^0$ and the cube holds $6r^2$

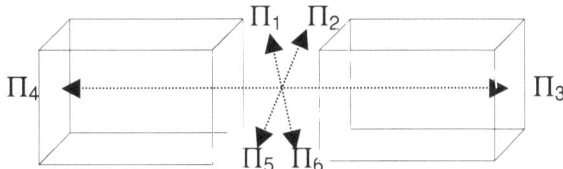

This is where the cosmos starts. This is where the element table has its root value. The Universe turns dimensional coming from being flat at Π^0 going to Πr^0 and then going 7/10 $(\Pi^6)/6$ =112. This is where the cosmos becomes a sphere (Π^6) in relation to the cube (outer space) having six sides (/6) in which the sphere turns (7/10). I explain many more such cosmic codes in **The Cosmic Code.**

Seeing our spinning top from the top, there are four quarters opposing each other and by that opposing one another. It is moreover the individual singularity in maintaining the major singularity, which sustains the governing singularity providing equilibrium in space-time.

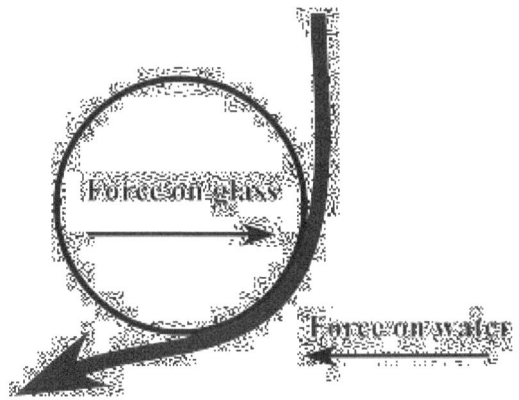

In Newtonian science there are more forces flying around than there are ghosts in an old cemetery. They even award forces to principles within the Coanda effect where the water receives its very own force or spook and the glass receives its own personal force or spook and all of this comes on top of the already existing four forces flying all over the show.

Newtonians, would you please believe me when I say there are no spooks, ghosts, fairies and forces and you can start to feel safe at night in the darkness with evil bats flying around in the darkness trying to haunt you because witchcraft is an old wives tale. Gravity is not the result of magic or unseen and unknown forces hiding within the atom and inexplicably trying to pull whatever it sees and then capture it by way of magic. The Newtonian way of looking at science is laughable when considering they are dreaming of something hiding within the star or the atom that grabs the next star billions of kilometres away through a vastness that offers nothing as a valid commodity and the grabbing hooks they call the graviton which hides somewhere in the mass of atom that then gives mass to the star that then gives gravity to the star with which it pulls and so the magic goes on.

The condition for the presence of this spot holding singularity the movement of Π^2 that is initiated by Π that forms a divide between eternity or that which is in the Universe that can never reduce because that is what can never start and singularity which forms eternity which is that part that can never end because that part can never increase.

The Earth spins through space at a value of (10/7 = 1.42) which then represents the interaction between space reducing and the Earth spinning. This is one very crucial part of the Coanda effect in establishing gravity. Secondly the Earth and all objects spin in a double motion where the moving straight represents the controlling singularity puts a spin directional diversion of 7° and the governing singularity puts in another directional diverting spin of 7° forming a total of (7+7) = 14. With the atmosphere (space-time) condensing by a value of 7 taken from the movement of the Earth and 10 as a value compressing the atmosphere the 10 forming a relation with 7 brings about space and time or solids (7) and air (10) mixing. I have already given the mathematical multiplication that then produces the movement of Π to form gravity or the numerical value of Π^2. In one dimension space became 10 and in that same dimension matter became seven. In order to separate matter (7) and space (10) through time (the spinning of matter) (7) in space (10) and space (10) spinning the matter (7) the following result came about through the application of the Roche principle $(\Pi/2)^2$. The idea of space compressing is what forms the principle that makes the Coanda effect viable.

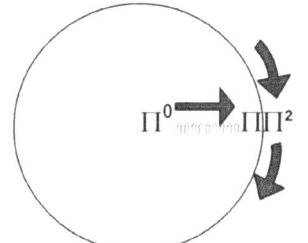

If not for the moving of Π to form gravity or Π^2 the value of Π^0 will run across the Universe as wide as it goes and hold singularity stretching into both in infinity as well as eternity with no breaking of the two time limits. As I have shown. The value of singularity holds no s[ace and with no parting of space the entire Universe will fall back into one point holding no space. The entire Universe will form one enormous Black Hole where time will disappear into and the Universe will again be no more because infinity then again locked eternity onto one spot holding time captured in one position. What forms the entirety called the Universe is the parting that Π brings into the Universe that gives credence to whatever has a measured meaning in the Universe.

Space mingles with time 10/7 and space separates from time 7/10. In this, the result is that matter forms an accumulating movement stretching into space allowing the solid to gain a time value. Time divided into space

$(10/7) \div (7/10) = 2{,}04$

$(10/7) = 1{,}4285 \div (7/10)\ 0{,}7 = 2{,}04$

Then also space moves towards matter compressing the cosmic gas to form cosmic liquid and this can become as dense as any solid could be. Light can cut through any material.

Space divided into time $(7/10) \div (10/7) = 0{,}49$

$(7/10)\ 0{,}7 \div (10/7)\ 1{,}4285 = 0{,}49$

Then space multiplied by time $(7/10) \times (10/7) = 2{,}04$

$(7/10)\ 0{,}7 \times (10/7)\ 1{,}4285 = 2{,}04$

That brings about that the Roche Principle
worked both ways (double)   1. $(7/10)\ 2{,}04 \times (\Pi/2)^2 = 5{,}033$
and with multiplication   2. $(7/10) \times (10/7) = 2{,}04\ (\Pi/2)^2 = 5{,}033$

Resulting in the combined value of $5{,}033 + 5{,}033 = 10{,}066$
On the other side the other combined value came to $0{,}49 + 0{,}49 = 0{,}98$
And the result from this product was
     $0{,}98 \times 10{,}66$

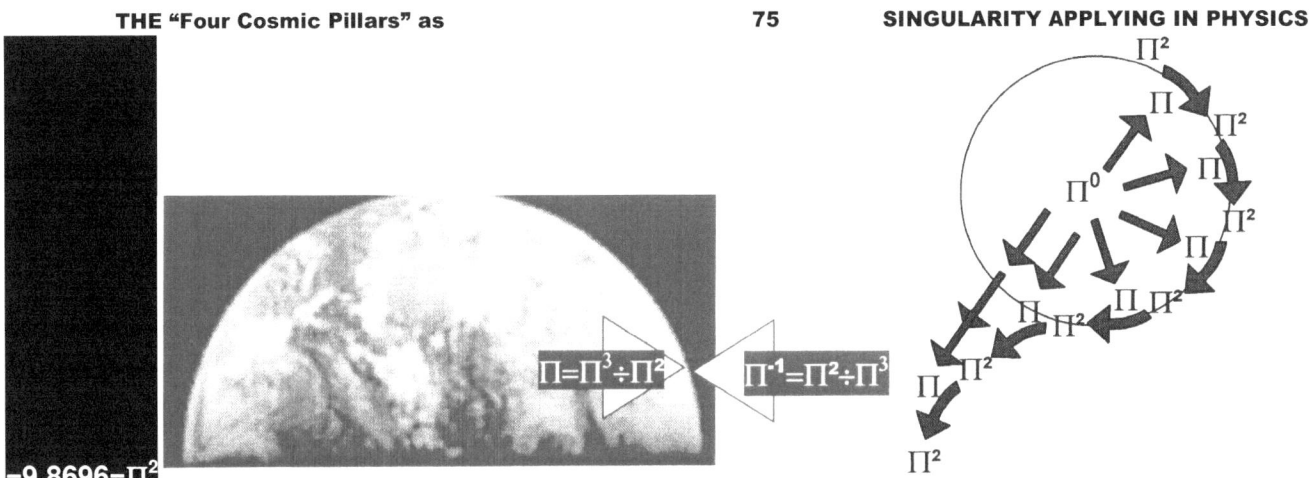

$$=9{,}8696=\Pi^2$$

$$\Pi=\Pi^3\div\Pi^2 \qquad \Pi^{-1}=\Pi^2\div\Pi^3$$

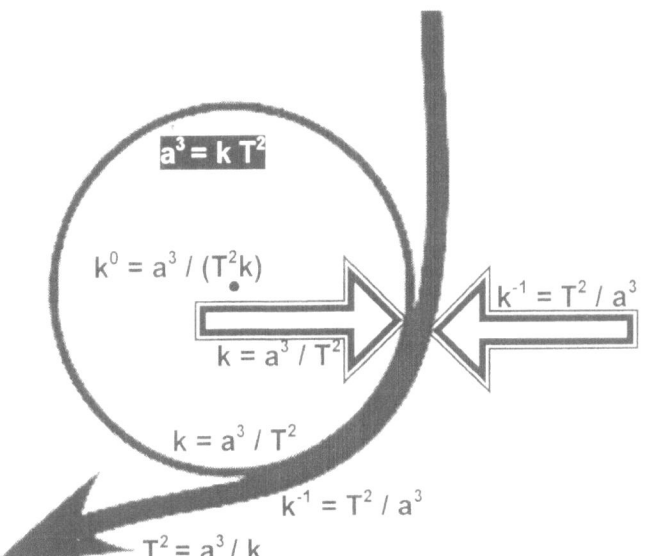

$$a^3 = k\,T^2$$

$$k^0 = a^3 / (T^2 k)$$

$$k = a^3 / T^2 \qquad k^{-1} = T^2 / a^3$$

$$k = a^3 / T^2$$

$$k^{-1} = T^2 / a^3$$

$$T^2 = a^3 / k$$

This shows that everything, controls everything and is everything is the <u>centralised</u> $k^0 = a^3 / (T^2 k)$ singularity that forms by movement $T^2 = a^3 / k$ of <u>space</u> $a^3 = k\,T^2$ <u>in relevancy</u> $k = a^3 / T^2$ both ways $k^{-1} = T^2 / a^3$ (Newton's 3rd law) thereof.

The only way one could explain the Coanda effect is by investigating the way Kepler interpreted the cosmos through the manner in which Kepler stated his formula. This is done by completely ignoring Newton's complete misconception into the work of Kepler. In this way it is also true that the only way one could explain the cosmos effectively is by investigating the way Kepler interpreted the cosmos through the manner in which Kepler stated his formula. This then too is done by completely ignoring Newton's complete misconception into the work of Kepler.

<u>This explains the Coanda effect and the Coanda effect is gravity and gravity "glues" the water to the glass in the same manner as how gravity "glues" air into the Earth or cosmic fluid onto the Sun.!</u>
This process happens to all spinning things and as much as it happens to a piston connected to a crankshaft, just as much this will happen to a atom spinning an electron in a similar manner as a the crankshaft is spinning holding a piston connected.

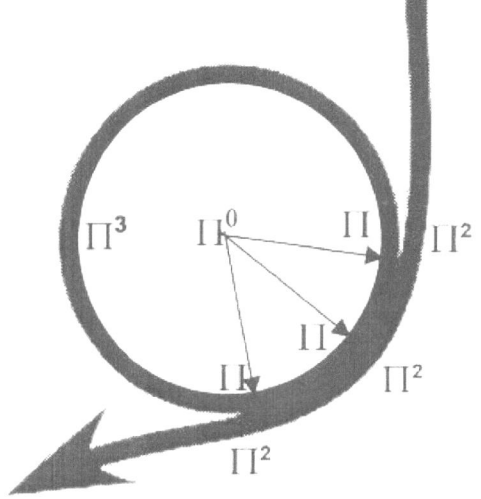

This concludes that $\Pi^3 = \Pi^2\Pi$, which when said in terms of mathematics seems very simple, but it explains an entire Universe formed by the movement thereof.

In order to initiate centre singularity Π^0, gravity Π^2 forming movement of space Π^3 is required. To initiate space Π^3 singularity Π extended has to be confirmed by movement Π^2. To secure space Π^3 movement thereof Π^2 must install space-time $\Pi^3 = \Pi^2\Pi$ that confirms singularity Π^0 forming solid Π^3 as well as singularity Π^0 confirming the liquid $\Pi^2\Pi$. This is what Kepler's formula brought to human knowledge and not once did I mention or indicate to mass being prominent or having importance or even being relevant.

This not only defines and confirms the Coanda effect, but it describes what happens with the Coanda effect and the Coanda effect is the result of the other three phenomena uniting and forming the culminating result. The **Coanda effect** is the coming together of the **Roche limit,** the **Titius Bode law** and the **Lagrangian system**, to form gravity not by mass but by movement of space filled by material in relation to fluid.

The Coanda effect #1
JL Naudin - 09-26-99

The Coanda effect #2
JL Naudin - 09-26-99

This proves that gravity is the Coanda effect and in another book I prove that the Coanda effect has its origins in Π forming a value and that value forms gravity.

THE LANGRANGIAN FIVE POINTS

In order to understand physics applying in cosmology I had to start by dissecting the set-up forming pi.

At this point I can introduce my theory on the ***Absolute Relevancy of Singularity*** At the point in the centre of the circle line must start. In the beginning when I explained the way I figured how the line start I said a lot of dots has to continue in order to form a line. It would be 1 + 1 + 1 etc. because the line must form by holding singularity after that point does mathematics begin but in the line that forms all factors holds 1. The lie can only form when all the points forming the line have the value of 1 being 1^0. In that conclusion one realises something must separate singularity from all other factors because singularity hosts all other factors but is by own initiative Π. Only when singularity meets the end value can the end value have Π where the final ring of the spinning circle forms Π. That will be the spot of origin forming the relevance in Π. That will hold the eternal spot…the smallest spot ever because all spots that ever can be were secured in a position in the centre of that spot that must continue as a line that forms. Because of the progress singularity follows from the single dimension singularity only allow mathematics a start at Π^0 progressing further too onto Π^0 and from there the line is born as $\Pi^0\Pi^0\Pi^0$ and to $\Pi^0\Pi^0\Pi^0\Pi^0$ etc. where Π^0 then may form the concept and value of r. But the line starts at $\Pi^0 = r^0$. This forms because cosmology is singularity based and the value is $\Pi\Pi^0$. This line $\Pi^0\Pi^0\Pi^0$ of singularity can only continue because every spinning atom preserves Π^0 in the very centre and since $\Pi^0 = \Pi^0 = \Pi^0$ the line is the same without finding conclusion except at the end where it forms mass at Π. At the point where Π forms, the

movement Π^2 of the circle defines the space Π^3 of the circle and it confirms the centre Π^0 of the circle through the rotation. Let's call this the solid forming or if you wish, let's call it Kepler's singularity. After that singularity forms a line $\Pi^0 = \Pi^0 = \Pi^0$ where this forms another line again as Newton stipulated it by $\frac{dJ}{dt} = 1^0$. Let's call that the liquid singularity or Newton's singularity and the relevance of singularity having a solid base compared to the singularity holding a liquid base comes about by the movement of gravity.

The cosmos started off with one dot so small eternity met infinity within. Then came one more dot holding eternal heat away from infinite cold, and this parting of heat from cold brought on another and another as the same repeated over and over. This parting of heat from cold was the start of the lot because this process is still ongoing today. I have an entire book written what this process involved as I interpret the interaction of the four phenomena into the process. This parting of the dot from the spot or parting of heat from cold then continued until there were a countless number of dots being as many as we now have that forms the cosmos we see. The accumulative size of the dots were the same size as just one dot was at the time because in the true Universe big and small plays no part of what holds value to the Universe we see. The dots were infinitely small and eternally big at the same time because size is a relevancy and without one the other has no size. So in the true perception, there is no difference in size.

What ever is started with the fact that there is no place or part in with which one may associate zero or nothing. There are no room for a number such as nothing. Every spot in the Universe has an eternal as well as an infinite value, which excludes nothing, or zero totally except in the understanding of the cosmos by Newtonians. Next to the one dot (infinitely close leaving no space to part the two points that represents singularity) one will find the next dot, and if nothing was a factor then that is precisely what one will find the two dots then are. The space parting the two points has no space but for the movement that forms Π where it is Π that grants the points individual characteristics. If it were not for the moving it was a non existing entity, taking up no space, and much more important, no time, therefore the dots are infinitely close to one another, being the same space, while at the same time being eternally big as much as infinitely small. If we as humans cannot find a manner in comprehending this notion, there can be no manner ever understanding the cosmos as much as the start to the cosmos.

When time began every dot was a Universe in its own and the accumulation was a Universe by merit of relevancy applying. The Earth in itself is a Universe as the moon is a Universe , as every atom is a Universe formed by Π^0 connecting Π and Π connecting Π^2 because rules applying on Earth do not apply on the moon and visa versa. When in the ocean another set of rules apply, therefore being in the sea places a body in another Universe . The number of Universal entities is still countless, just as much as it was in the beginning.

Every position in the Universe either holds singularity in a form, or relates to singularity because every spot there is, is also singularity. There can be no position unrelated to singularity therefore every aspect of the cosmos is space-time in various forms under the provision of singularity connecting. Matter cannot be if not being surrounding by singularity and secluding singularity.

Singularity is as close as any spot can ever come to zero BUT IT CANNOT EVER BE ZERO. From singularity diverts space-time and there cannot be space without time as much as there cannot be time without space, not withstanding the size of space or duration of time.

Every dot insignificantly small as it may be, is a part of another Universe as much as it is part of the accumulative Universe and every dot in the infinity holds singularity, which we translate as " nothing" being "darkness or being beyond the noticeable". There cannot be "nothing" just as much as there cannot be "darkness". There cannot be something big or small, but when it is put into relevancy of perception, and then the relativity of perception becomes the question. There cannot be hot as much as there cannot be cold, big or small, far or near, bright or dark because every "boundary" mentioned is just a form of development processes we humans attach to something we will never fathom. The Sun FREEZES hydrogen to a liquid at six and a half thousand degrees Celsius and Universe boils over in the form of the Hubble constant at the temperature (we presume from our vantage point) at minus 273 degrees C. If we Humans cannot or will not abandon our human perception and our manly perspective, we may as well return to astrology for all its worth. Even the atheist places life in the pivot of the Universe while in reality life is on one very small insignificant and unimportant planet spinning around an even less prominent underdeveloped star.

Space-time is a four dimensional position of the Universe where the position of an object is specified by three coordinates in space and one position in time.

With singularity placed in infinity within the centre of every rotating object every atom and its relation to its surroundings including other atoms form space-time diverting from the point holding singularity as far as rotation goes because every object holds three relative positions in as far as where it was, where it is and where it will be in relation to singularity providing time. I elaborate on this else where.

From singularity Π^0 the line runs to both sides forming the edges of singularity Π forming the border through singularity moving Π^2. All aspects of the cosmos hold two halves in four quarters while rotating. One the left side (1) there are two directional changes (2) and on the right side (1+1) there are two more directional changes (4). Each of the halves and all of the quarters are in direct opposition to each other as much as to one another. Within rotating 180° the one point will represent all it had just 180° before in the very opposing way. This is why the Earth's magnetic fields swap every ($\Pi \times 10^3$) year and ice ages change which is an ice desert into heated desert drought cycles and back again. From any point all space is moving from one side of the Universe to the other side of the Universe in quarterly displacement. The time it takes the movement will be the movement from the starting point at a value of Π^0 to an ending point holding that moment of infinity to a value in eternity therefore bringing the square of Π being Π^2 which then forms the value of time. Because time is the movement bringing about the change in quarters from any given point to any other given point nothing can be in two places of the Universe simultaneously

The following sketches present the viability of the
Lagrangian Points

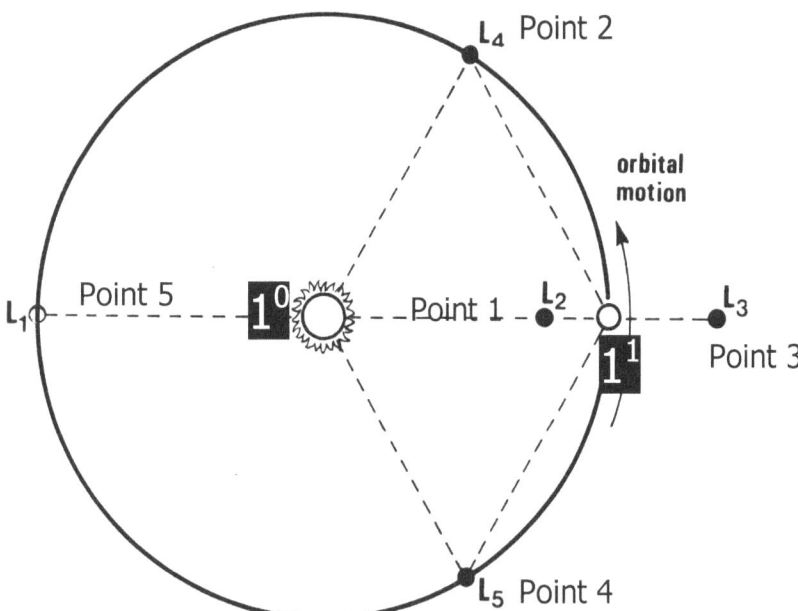

There are five sides connecting to the circle that has seven sides. This personifies the space part where the Bode law personifies the material part.

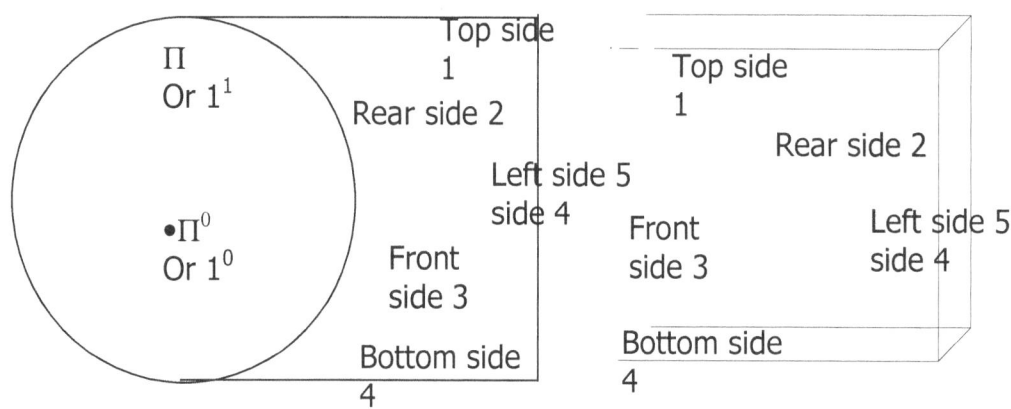

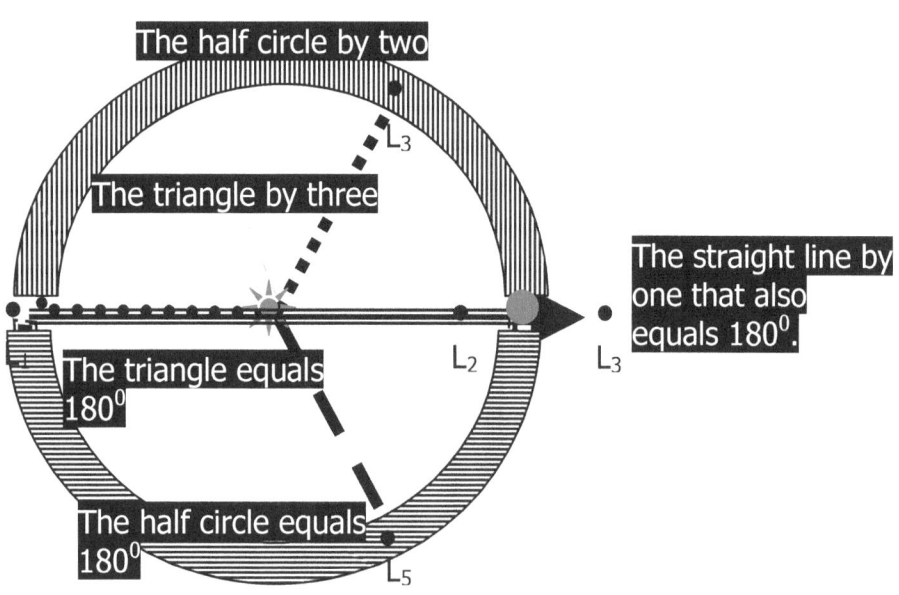

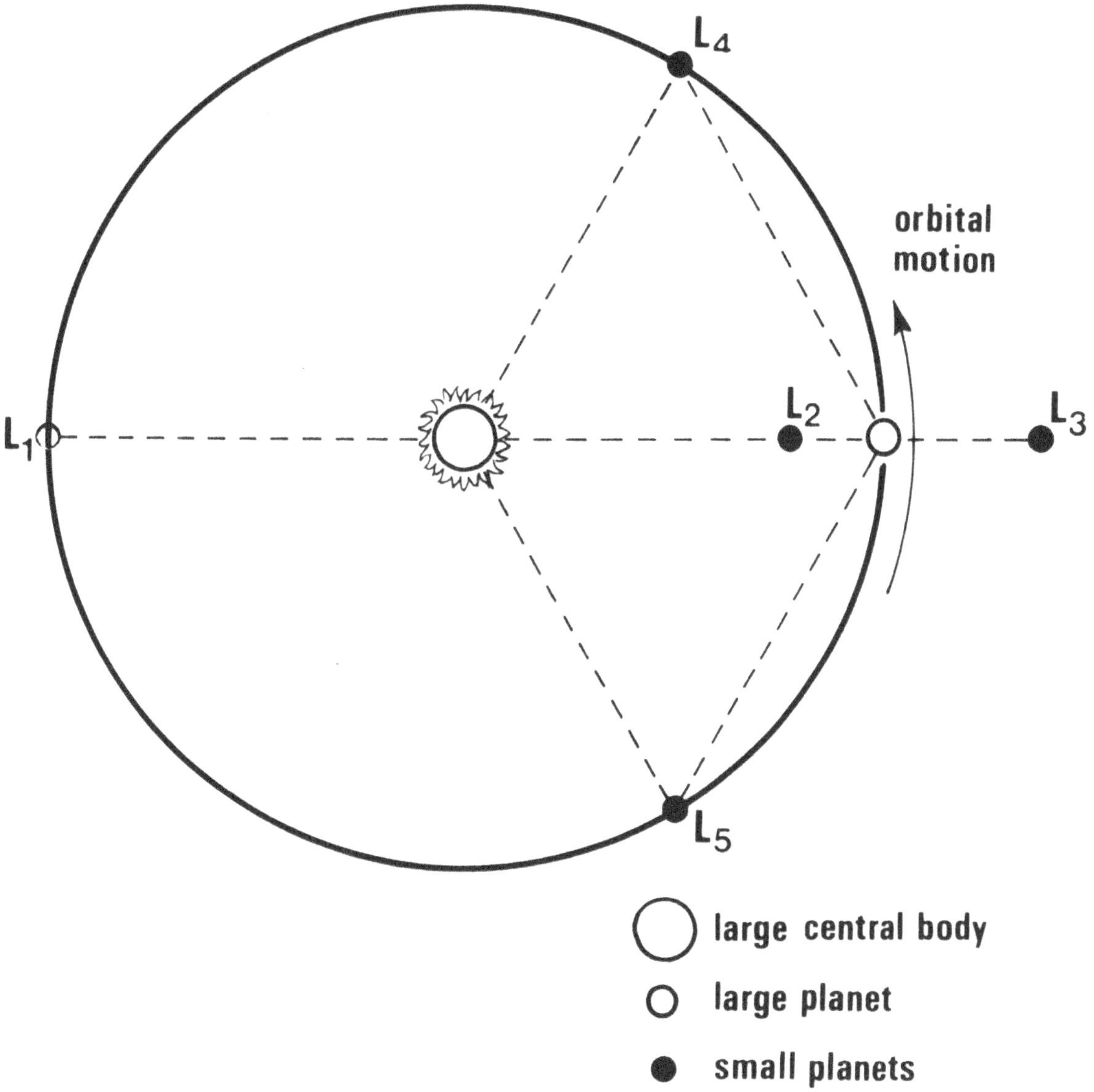

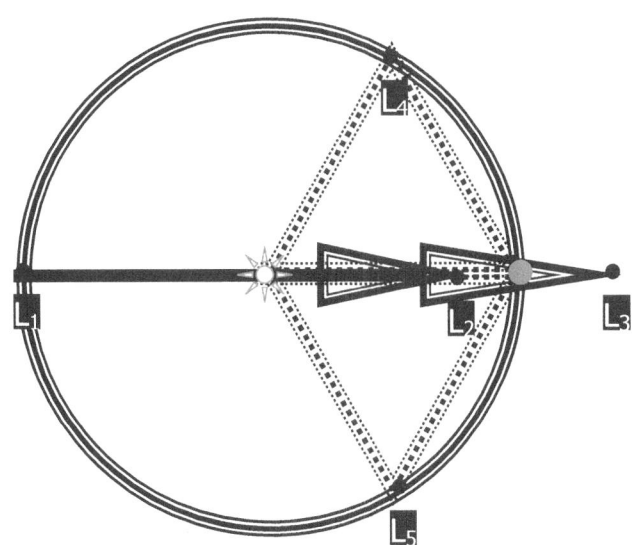

L4
orbital
motion
L1
L2
L3
L5
large central body
large planet
small planets

L4
L1
L2
L3
L5

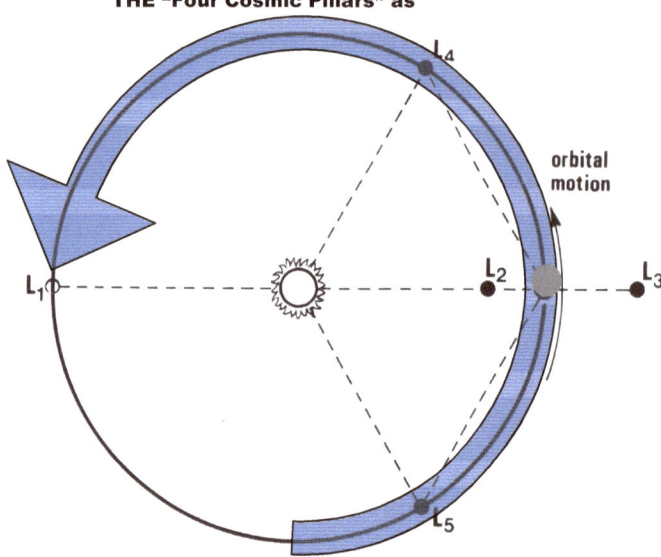

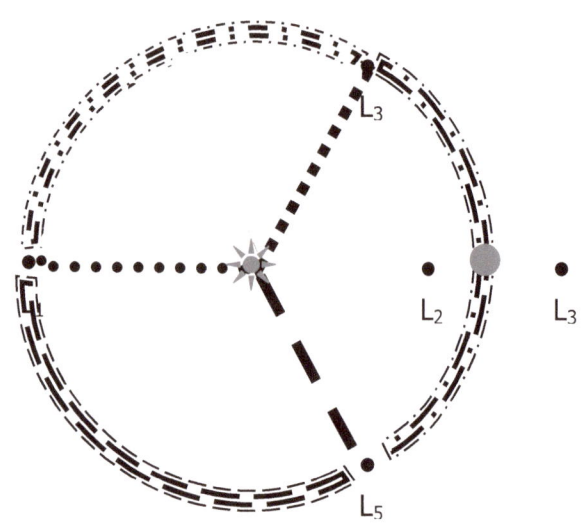

THE LAGRANGIAN SYSTEM.

LAGRANGIAN POINT:
The Lagrangian points
are five equilibrium points
in the orbit of one body
around another, such
as a planet around the Sun

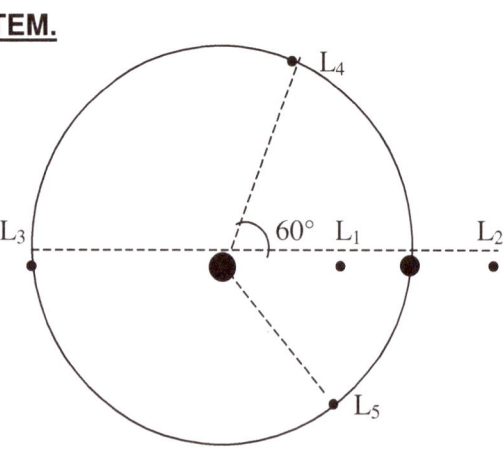

When going into any one of the four cosmic pillars one has to keep in mind constantly that what applies, applies at a point where singularity turns to space. It happens where mathematics change numerical from holding only form to becoming valid digits If anything was part of the Universe only once that something is part of the Universe forever because then that something has no where to go but to remain in the Universe . At this moment time forms as singularity going into forming space as time moves on and space becomes the visual history of time gone by. But not even time can disappear out of the Universe. Time leaves space behind as the history of time. Newtonians think of mathematics as godly forms by thinking it is mathematics that formed the Universe while the reality is that it is not mathematics that formed the Universe but it is the Universe that formed mathematics. It is clear that using form in the cosmos at the beginning became a principle before mathematics or numbers became valid. This is what the Lagrangian points prove. A triangle, a straight line and a half circle doesn't even approach to be the same in form and yet all three are the same in directional value as 180°. This proves that before mathematics was valid there were other values already in palce and from that principle applying a numerical Universe came about. The overbearing reason Newtonians why science can't to an era before the Big Bang is that they do not understand this issue of the Universe being there before mathematics was there. There is a time that proceeded the time Newtonians get stuck at which is when space as heat came about which they call the Big Bang. There was a time when form got to form a Universe when space was still a thought that stretched far into the future. Mathematics proves this as much as that mathematics proves that there was a time when directions were the only valid mathematics used by the Universe. By the way, the name the Big Bang is as silly an idea as the Newtonian idea that mass forms gravity because at the time there was nothing big and sound was just a thought into the future.

It starts as follows:

No line can start at zero because having a starting point of zero there is no line (0 X by what ever reduces whatever to zero). The starting point has to be infinity the shortest any line can be leading to eternity the longest any line can ever be. By having infinity there then has to be a VERTUALL ZERO (not zero) and from that point the rest of the line must start running the other way.

At a point where singularity attaches to space in time the value of the straight lien is equal to the half circle, which is equal to the triangle. In the book **An Open Letter one Gravity** I go into detail as to explain why this occurs but that would take up more space than this entire book permits. To understand the Lagrangian system one has to come to terms with time and what time is. The Newtonian surmising that time can be one and then being 1 time can stand still as depicted in the formula $\mathbf{T = 1 -\sqrt{C^2 - V^2}}$ the is a folly as most other ideas seems to be. Time is always carried by the value of three and that is the responsibility of the electron. Time is in the past carried on to the future while moving onto the future.

When one slows the electron down to a point of near standing still one can observe the electron in two positions. That is not the case. What one sees is the electron departing from the past (visible point) going onto the present (point while observing) being in the position of observation while at point three (visible point) departing to the future. Let's slow this down somewhat while remaining relevant to the process. When I see a Super Nova event, the event that is taking place was in my past and therefore it forms part of my past. The light streaming towards me is confirming my past as that is what happened. The light that I see with which I observe the event is what confirms my present as the light is streaming to me by bringing the present. Furthermore the light streaming towards me is confirming my future since the light coming from my past still has to reach me in the future which is also the light that will be at the point where I am in the future and therefore while coming from the past and confirming my presence the light coming from the past is also confirming at the same time my future that I have. That makes me relevant to the Super Nova occurring and that gives meaning to the Super Nova as it catties the present to the future. Therefore, The light is confirming my past (1) (the event that took place) while witnessing the event that took palce long ago my witnessing of the event is also confirming and establishes my present (2) (reaffirming what happened by putting me in a time position in relevance to the past) while the light running from my past confirms my future (3) by determining in relation to the past that I have a present since my past becomes my future while waiting for the light from the past to arrive at the present. This account is only the tip of the iceberg because when we start to dissect how light uses time to displace space in order to move the event onto and into the future to where I am the argument really gets intensive and it takes many pages of explaining the complicated issues in hand. This has no bearing on the Lagrangian system but only serves to point out that there is a position where time places space into the Universe in order to have space serve as the history of time. His is the point where the four cosmic pillars serve time in s[ace as to use time to form space. This singularity, which is a point time, places being without space to form a relevancy with space by afterwards becoming space. Singularity being time that is going into space by forming Π is the "white hole" that is in contrast with "the Black Hole" that every one was forever looking for.

The Lagrangian system is part of the four pillars on which time or gravity rests. The Lagrangian system is Kepler's formula put into principle. It is the manifestation of $\mathbf{a^3 = T^2 k}$. It is the movement ($\mathbf{T^2}$) forming space ($\mathbf{a^3}$) in relation to time ($\mathbf{k}$) related to singularity ($\mathbf{k^0}$) where singularity is excluding space-time ($\mathbf{k = a^3 / T^2}$), including space-time ($a^3=kT^2$) and secluding ($T^2= a^3/ k$) space-time by motion of space-time in time forming a negative or opposing curve ($k^{-1} =T^2 / a^3$). One that cannot see this principle in practise as singularity improvise space by replicating time, such a person must be blind or a Newtonian, which makes no difference either way, it is still the same person seeing only mass and nothing.

The Lagrangian points system is the manifestation of time forming space-time by leaving space as delayed or historical time. I am not going to explain this for there is other books that detail what I say. Every time space ends in four, the very next point confirming singularity will be point five where singularity then provides point six and seven as a non movable line. It is the principle of how time rolls into space and the Lagrangian system provides the full range as to how time interprets space as the delayed form of time. The space forms five points (four in time plus one in continuing the alignment) and from that the line shows that the seven points positioning singularity forms the sphere. Time moves through the four sectors of the circle and the very next point in which space confirms s a new point (point five) from where space will grow into a new circle or cycle.

The Lagrangian system is working as the growing part of the Coanda effect where the liquid holds singularity Π^0 extended $\Pi^0\Rightarrow \Pi$ in relation to forming a liquid in motion Π^2 in relation with the solid that forms as Π^3 confirming the liquid borders. Where $\Pi^0\Rightarrow \Pi$ in relation to forming a liquid in motion Π^2 to form the attachment of $\Pi^0\Rightarrow \Pi$ is the indicator of the attaching of fifth point. The liquid holds solids as markers in five locations and the liquid spins around the solid as the solid confirms the liquid by singularity extending to form the border confirming space-time. The Lagrangian system is one part in three other that forms the Coanda effect as the cosmos interprets gravity forming and the Lagrangian system is the improvisation for gravity by way of the Coanda effect.

But most of all, the Lagrangian system delivers every proof of how singularity forms space-time. It shows how singularity Π^0 places the relevancy of Π in relation to gravity moving Π^2 and in this movement the space forming contains the Universe that singularity provides to secure and define that space-time it preserves asΠ^3.

The Lagrangian system proves the singularity places the straight line (180^0) equal to the half circle, each being (180^0) and that is put in equal relation to three triangles, each being (180^0). The Lagrangian system indicates the manner which singularity uses to put in place the characteristics of space-time relating to singularity. It is the straight-line (180^0) connecting the (two) half circle(s) (180^0) to the (thee) triangle(s) (180^0).

But moreover it is the manifestation of the atom as singularity improvises the connecting of space-time contraction. Every star is doubling as a cosmic atom and every star is just as much an atom as being the accumulative product of all the atoms combined.

Singularity can only meet space by measure of a straight line that is forming is running to both sides forming the edges Π of singularity starting at Π^0 the border Π of singularity Π^0. This means that all aspects of the cosmos hold two halves in four quarters. Each of the halves is formed by Π meeting space on either side of the Universe (two equal circles forming by the measure of Π). The movement forming Π holds a relation to a position (point) that forms (one digit) in relation to the other side having two reference point By having the three points forming three quarters that hold two halves that are separated by one straight line five pints in all form. In the full rotation that comes about as Π moves to form Π^2 of the quarters form that is in direct opposition to each other as much as to one another. The four quarters hold a value of fifty each where in the Pythagoras triangle $7^2 + 1^2$ forms the fifty. Since space holds a value of 10 in relation to the 7 that gravity brings about, the dividing of 10 into 50 puts 5 places in a dimensional alliance with the 10 of space to form the fifty that From any point all space is moving from one side of the Universe to the other side of the Universe in quarterly displacement.

Then forming Π as a full rotating circle the 4 directional changes holds 5 points each that positions 20 points in the full circle. This is how the value of Π or 3.14159 comes about. Having singularity in development (.991 or put into context with the circle Π - 3 = 0.1416 and 0.1416 x 7 = .991 which means that singularity expanding into space holds 1.) On the other side the three movement of gravity forms 7 x 3 = 21 points and with singularity growing as time another 0.991 adds to 3.14159 or Π. That brings about the proof that Π is the way gravity forms a continues decline in the density of outer space as well as an increase in volumetric occupation of space within matter and the fact of gravity is vested in the value Π carries. In **The Veracity of Gravity** there is a lot more explaining as to how this comes about. The time it takes the movement will be the movement from the starting point at a value of Π to an ending point holding that moment of infinity to a value in eternity therefore bringing the square of Π^2 to the value of time. Because time is the movement bringing about the change in quarters from any given point to any other given point nothing can be in two places of the Universe simultaneously.

LAGRANGIAN POINT:

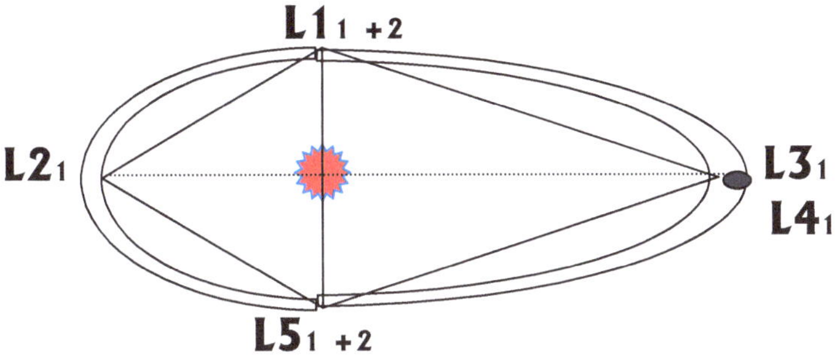

L1₁ ₊₂

L2₁   **L3**₁
                 L4₁

L5₁ ₊₂

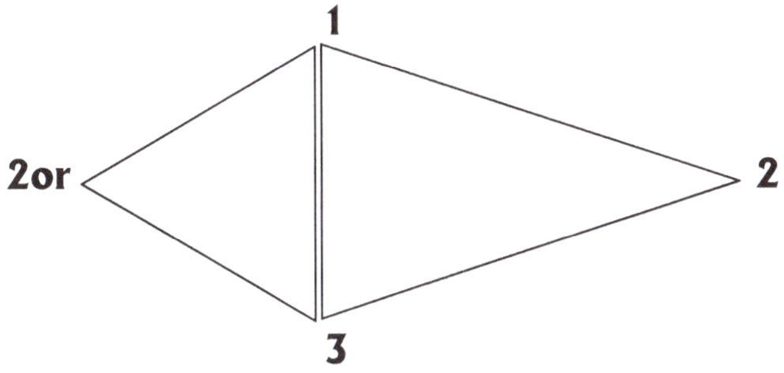

1

2or

2 = 3

3

Singularity presenting the triangle with 3 markers each only on one side of singularity.

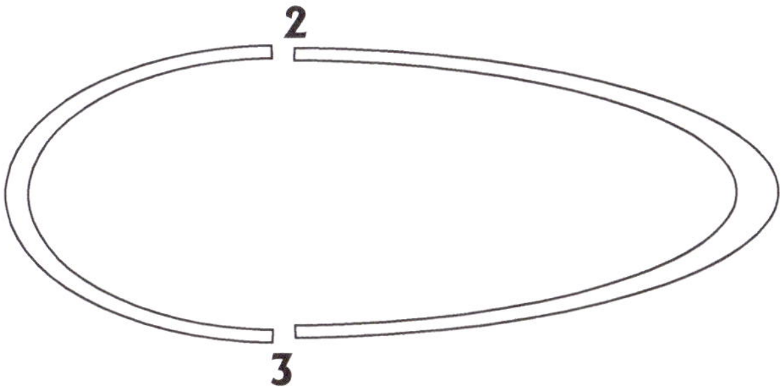

2

3

= 2

Singularity presenting the half circle with two marking points

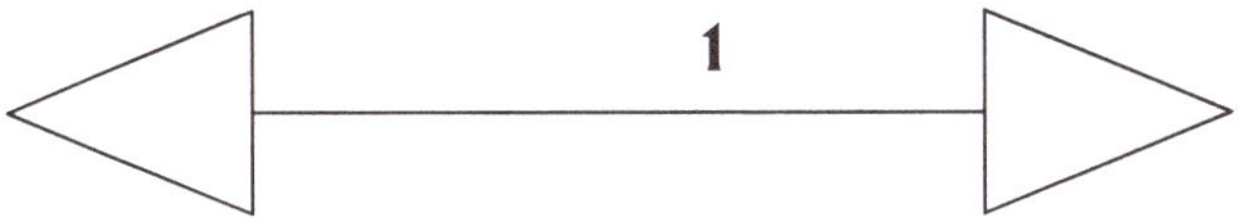

1

= 1

Singularity presenting the straight line with one point forming part of the circle

Triangle holds three, the circle holds two and the line also holds two points it shares with the circle by bringing a dividing half sharing points as singularity infinite.

It all forms part of singularity as a unit by three , two and one holding five in total and with space being the result from matter dismissing Π to favour r space must either join matter by becoming Π and dismissing r or maintain r and hold a maximum of five points to singularity at greatest value. With the Universe always in division by singularity the singularity holding seven position to Π will relate to the two singularities affecting the position of a cosmic atom. That will form as double points to space where five then multiplies with the two aspects of singularity divide and form the value of 10.

LAGRANGIAN POINTS in the LINE:

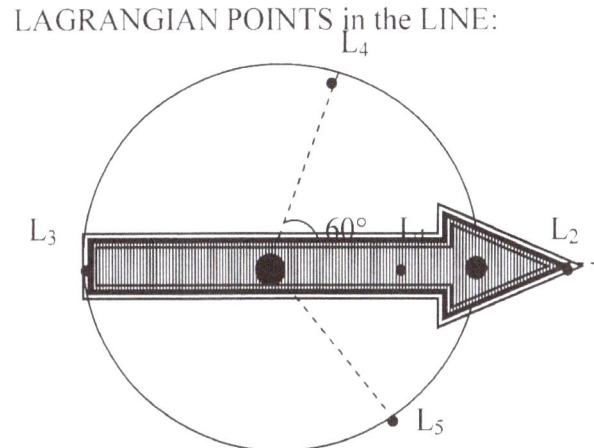

LAGRANGIAN POINTS in the TRIANGLES: LAGRANGIAN POINTS in the CIRCLES:

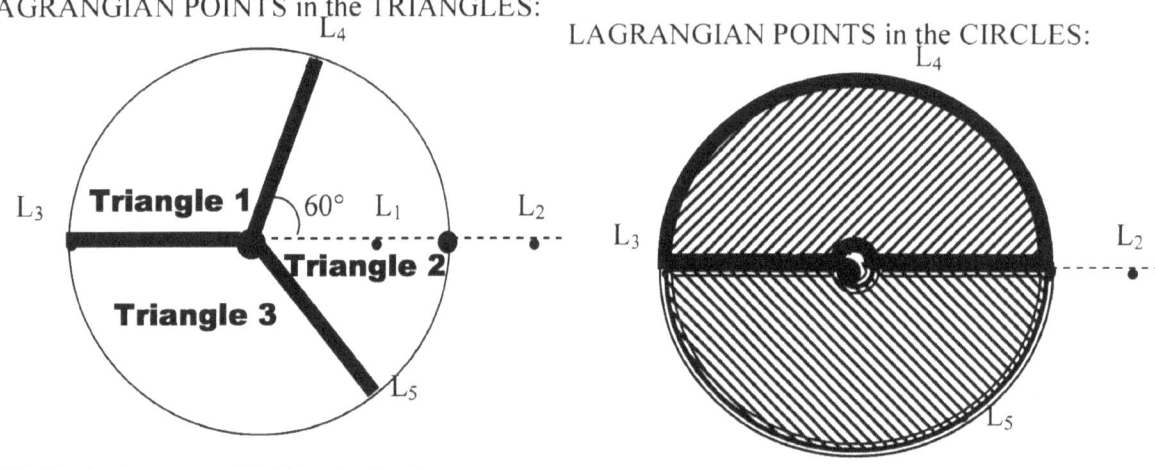

1 Half circle	= 180°	L₃ L₄ L₅
2 Triangle 1	= 180°	L₃ L₄ L₅
3 Triangle 2	= 180°	L₃ L₄ L₅
4 Straight Line	= 180°	Singularity
5 Double Circle	= 720°	Sphere

From singularity extending into space there comes three values each holding 180⁰ and this fact science is familiar with. The values predate mathematical numerical and formula values and prove the most basic of mathematics was in place long before numerical values start applying. From this the numerical formation came about and I wrote a book dealing with such a matter of development. As everything grows from singularity the straight line is always a potential triangle with on side apparent and the other side in infinity.

There is no zero from where a line can start or grow and because of the absence of such a point mathematics brought about a diversion to escape the zero mark not existing. This abolishes the century-old idea that a line springs from zero or that a graph implicates zero as a divide. If the straight line did cross zero it would not be one line but it would be no line since the one line will discontinue cancelling the line at one point. All lines have to have a start and end at a single point without space therefore no line can be half a line. Therefore every line that forms is also a circle just as much as the circle includes a straight line to represent space just as Kepler said it does in $a^3 = T^2 k$

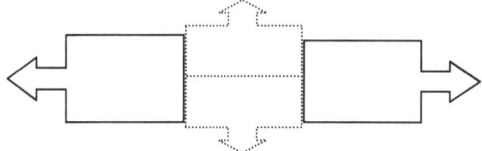

In order to overcome such a problem the straight line holds another line Π on both sides Π² as a point in infinity Π⁰ to half the line as to enable the line diverting from 1⁰ and grow to Π.

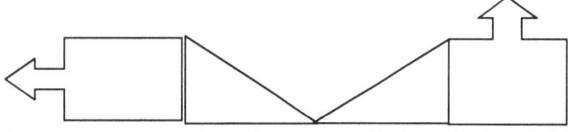

Because each line represents the other side of singularity dividing singularity by half the square of such a halves represent singularity two the half of a circle thus bringing total of the two halves would match the other half of singularity in half the circle. This is very important when considering the way that light travels in time through space.

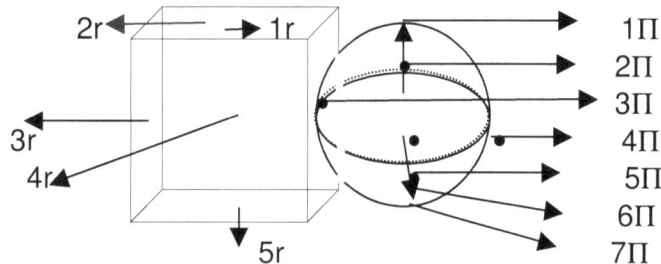

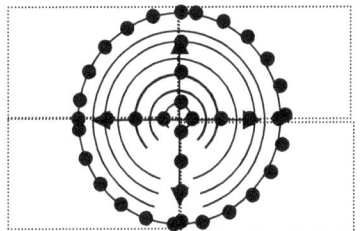

From space the cube holds the value of 5 times four quarters in relation to singularity forming the four five points of the square in the cube.

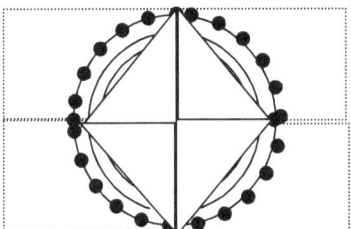

From singularity holding the relevancy the five sides in the cube as a square holds four triangles to two circles.

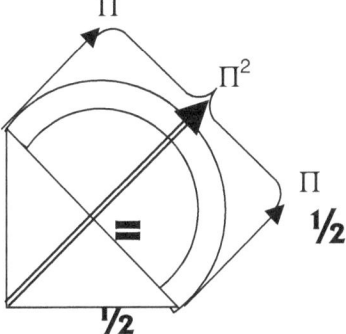

$\frac{1}{2} + \frac{1}{2} = \sqrt{1}$

$360^0 / 5 = 72 / \prod = 22.91$ which is 1 more than 21.991

What this proves is that when the circle is completed singularity bringing the future (10) is already in place because $360^0 /5 = 72/ \prod = 22.91$ shows a value one point more than $\prod$ because $\prod$ is 21.991 in relation to 7.

1 is singularity on the one side of the Universe and 0.991 is singularity on the other side of the divide.
10 from singularity on the one side of the Universe and
10 from singularity on the other side bringing about the
$\prod$ that holds 21.91 to 7
Since the sphere is double the circle and half the circle represents singularity by the square, half the square of the triangle is a straight line diverting singularity the law of Pythagoras is valid.

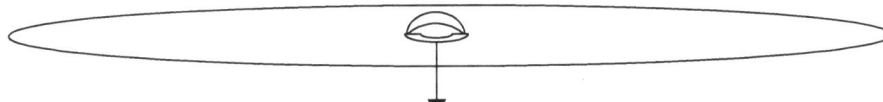

The divide bringing about the two sides of the Universe where the one (1) to singularity depicts the one side of the Universe and the other side depicts the other side holding space from singularity (.991) bringing about the singularity value of 1.91

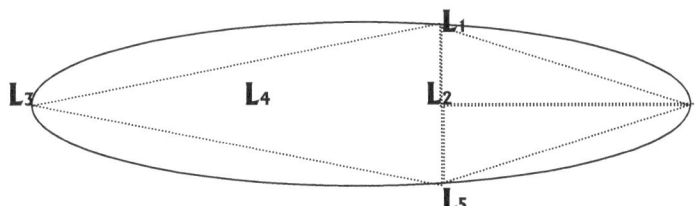

Dividing the four fives singularity holds a centre line (1.991) with one on one side and 0.991 on the other side but since it is a space relating to a sphere only one of the quarters on either side of the divide relates to a specific therefore unlike the sphere where the full value of $\prod$ relates to four fives, bringing about $\prod$ as the dominant the space separating the sphere from the points in space holds a combined value of one cube in line with the divide singularity supplies having five points.

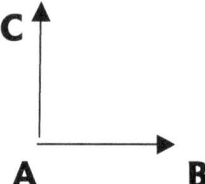

It takes any line time to relate to space and only nothing is instantly and nothing is what it is it is the absence of anything, therefore moving the line through the space it covers is the square of time in as much as Π^2. The time factor stands in all cases in relation to the space factor in the square because time being single dimensional develop space as three-dimensions while the space is in the sphere that spins in the cube and that is a law of cosmology. To that reason is why I reintroduced Kepler's formula as to explain the cosmos by the formula $a^3 \div T^2\, k = 1$ because this is the way the cosmos gave information to Kepler. Using $a^3 \div T^2\, k$ puts a relevancy between time $T^2\, k$ in the line and space a^3 to the line.

Since the triangle in singularity are on both sides of the divide of singularity and the circle holds the time aspect relating to space in the square, therefore the triangle then must relate to space in a square in order not to duplex singularity in the divide.

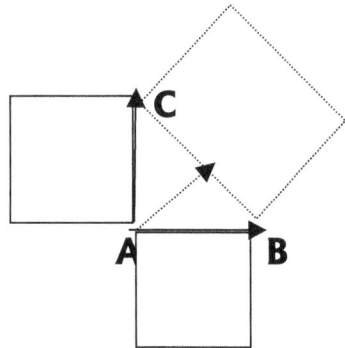

The time affecting the space of AC will relate equal to the time the line AB relates to time and space and where time is always in the square the lines will be the square of the triangle forming in relation to the square existing in the total of the time to the space relation forming between the lines.

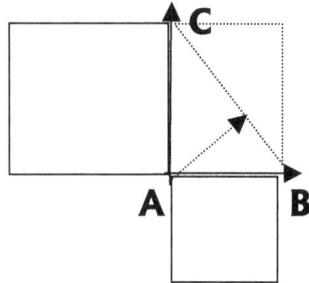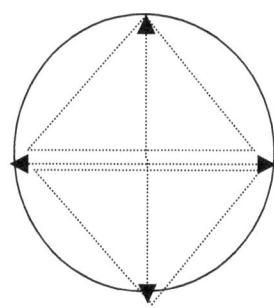

Since the triangle forms on both sides of the divide and all things concerning singularity is in duplication the double triangle will be the square. In the same way the circle represents both sides of the divide of singularity that forms and then also has the value of 2 X 180^0 as does the double triangle triangles. That brings about that Π relates to the square Π^2 to form three-dimensional space and from this fact mathematics can substitute Π by using r^0. However that is not the case in the dimensional aspect and as mathematics gets its queue from the cosmos and not the other way around the substation may apply to singularity as space in the sphere but not as space as the sphere. Heat is concentrated space and space is expanded heat. Gravity and electricity is the very same thing where electricity is a concentration of heat demolishing space in a very specific location and gravity is the concentration of heat in a less dynamic form but acting in a much broader space. In both instances it is the polarized motion of iron $_{56}$ linking time directly to singularity through the conducting of heat surrounding elements. To this effect I wish to point out that no element is either a liquid, a gas or a solid as all elements are all three forms and it is only the state of the relation that apply to an element at a very specific position in space-time that will allow the element to act in either of the conditions that the heat or space which is the same thing will allow. Gravity is the dimensional destructing of space to the concentrating of heat in that space by increasing the time duration through the Titius Bode principle or when matter holds less space to the normal allowing of matter occupying space, matter will produce the Roche principle in guarding its individual singularity of the mutual singularity between objects.

This principle also is the only difference of notoriety between electricity on Earth and gravity on earth. On Earth Π has a very slightly higher value than space in as much as space is 3 and the atmosphere is Π. In a cosmic midget as the Sun are the relevancy changes considerably and the space to atmosphere can be 10 Π. There is no chance of generating electricity in the atmosphere of the Sun because the atmosphere of the Sun is electricity in as much as the gravity being 10 Π to the Π of the earth. That explains the fact that the Sun liquefies heat to a watery substance. With the heat in a liquid the Sun becomes a sea of heat.

By matter applying Π as the reference there is little man can do to change that. In the case of electricity using r to form the value C we can change that because as r relate to space and we are part of space as space above the Earth in the neutron zone we can change the space holding r as value. When looking at the Sun applying gravity and relating that we see to what we find in electricity there is hardly any thing to recognise a similarity. But when in view of dimensional dynamics it is the same thing because in the Sun even r holding heat becomes Π holding matter as heat becomes liquid and that stands in between matter and gas. With this view in mind it would be worthwhile to have another look at the way we see how creation started and bring heat in as related to r and matter being related to Π.

Gravity is the transformation of space to heat in one specific dimension changing that particular dimension in relation to the other five dimensions. That brings the reason why the Lagrangian system can only allow five positions and allowing any more will destroy any form of dimensional implication between object relating to one another while sharing space occupying in time duration. The Big Bang had its massive motion brought on by first implementing the Roche factor of $(\Pi / 2)^2$ after which when matter had a larger claim to space and space broadened the Roche factor adjusted to $\Pi 2 / 2$ and then implementing the Titius Bode principle very much later on toΠ.

I also prove that gravity is the result of four cosmic phenomena interacting to form the value of Π which by movement becomes the value of gravity Π^2 and gravity is equal to cosmic time applying in other books also available form LULU.

1) *Absolute Relevancy of Singularity in Relation with Applying Physics*

1) The location, the position and the value of **singularity** as a factor forming space-time
2) Finding **space-time** by dissecting Kepler's formula in relation to valuing singularity
3) Finding space-time, **proving space-time** and **aligning space-time** with gravity.
This is part of the ***Absolute Relevancy of Singularity in Relation with Applying Physics***

2) *Absolute Relevancy of Singularity in Relation the Four Phenomena*

4) Finding the **Roche limit**, and explaining the resulting of a law coming about from singularity.
5) Finding the **Lagrangian system**, how and why that becomes the building form of the Universe .
6) Finding the **Titius Bode law** and I show mathematically how gravity comes about from that
7) Finding the **Coanda effect** and the producing of gravity through reproducing space-time.
As part of the ***Absolute Relevancy of Singularity in Relation the Four Phenomena***

3) *Absolute Relevancy of Singularity by Explaining the "Sound Barrier"*

9) Proving the phenomenon known as the "**sound barrier**" by proving it **is gravity** generated **by motion** in space becoming independent where motion creates independence. Breaking the sound barrier is the motion in space duplicating space by crossing over gravity borders. It is $a^3 = kT^2$ where $(k \leq T^2)$ or $(k > T^2)$
As part of the ***Absolute Relevancy of Singularity by Explaining the "Sound Barrier"***

4) *Absolute Relevancy of Singularity in Relation with Explaining the Cosmic Code.*

4) Finding the **working principals** behind and manifesting **of gravity** as a cosmic occurrence and introduce it as part of the ***Absolute Relevancy of Singularity in Relation with Explaining the Cosmic Code.***

Books Confirming This presentation

How the Solar System Forms: An Academic Presentation by Peet (P.S.J.) Schutte
ISBN-13: 978-1523217021 (CreateSpace-Assigned)
ISBN-10: 1523217022

A Cosmic Birth as an Academic Presentation Book 1 by Peet (P.S.J.) Schutte
ISBN-13: 978-1517066970 (CreateSpace-Assigned)
ISBN-10: 1517066972

A Cosmic Birth...as a Special Presentation Book 2 by Peet (P.S.J.) Schutte
ISBN-13: 978-1517525460 (CreateSpace-Assigned)
ISBN-10: 1517525462

An Academic Introducing to The Titius Bode Law Book 1 by (P.S.J.) Peet Schutte
ISBN-13: 978-1507845851 (CreateSpace-Assigned)
ISBN-10: 1507845855

An Academic Introducing to The Titius Bode Law Book 2 by Peet (P.S.J.) Schutte
ISBN-13: 978-1507853788 (CreateSpace-Assigned)
ISBN-10: 1507853785

An Academic Introducing to The Titius Bode Law Book 3 by Peet (P.S.J.) Schutte
ISBN-13: 978-1505874884 (CreateSpace-Assigned)
ISBN-10: 1505874882

How the Solar System Forms: a Pre- Script by Peet (P.S.J.) Schutte
ISBN-13: 978-1503023895 (CreateSpace-Assigned)
ISBN-10: 1503023893

Relevant applying literature Go to Google Amazon.com: Peet Schutte: Books
http://www.amazon.com/s?ie=UTF8&page=1&rh=n%3A283155%2Cp_27%3APeet%20Schutte.
Oxford dictionary of Astronomy web site naturescosmicconcept

The Following books are all available from CreateSpace web site.
The Absolute Relevance of Singularity The Journal
The Absolute Relevance of Singularity The Unpublished Article
The Absolute Relevance of Singularity The Dissertation
The Absolute Relevance of Singularity in terms of Newton Book 0
The Absolute Relevance of Singularity in terms of Cosmic Physics Book 1
The Absolute Relevance of Singularity in terms of The Sound Barrier Book 2
The Absolute Relevance of Singularity in terms of The Four Cosmic Phenomena Book 3
The Absolute Relevance of Singularity in terms of The Cosmic Code Book 4
The Absolute Relevance of Singularity in terms of Life Book 5
The Absolute Relevance of Singularity in terms of Investigating Kepler Book 6
The Absolute Relevance of Singularity in terms of The Thesis Book 7
The Absolute Relevance of Singularity in terms of The Cosmic Creation Book 8

peet@naturescosmicconcept.co.za   mail.naturescosmicconcept.co.za

www.ingramcontent.com/pod-product-compliance
Lightning Source LLC
Chambersburg PA
CBHW050733180526
45159CB00003B/1210